Holger van Bargen

Some Asymptotic Properties of Stochastic Flows

Holger van Bargen

Some Asymptotic Properties of Stochastic Flows

Isotropic Brownian Flows and beyond

Südwestdeutscher Verlag für Hochschulschriften

Impressum/Imprint (nur für Deutschland/ only for Germany)
Bibliografische Information der Deutschen Nationalbibliothek: Die Deutsche Nationalbibliothek verzeichnet diese Publikation in der Deutschen Nationalbibliografie; detaillierte bibliografische Daten sind im Internet über http://dnb.d-nb.de abrufbar.

Alle in diesem Buch genannten Marken und Produktnamen unterliegen warenzeichen-, marken- oder patentrechtlichem Schutz bzw. sind Warenzeichen oder eingetragene Warenzeichen der jeweiligen Inhaber. Die Wiedergabe von Marken, Produktnamen, Gebrauchsnamen, Handelsnamen, Warenbezeichnungen u.s.w. in diesem Werk berechtigt auch ohne besondere Kennzeichnung nicht zu der Annahme, dass solche Namen im Sinne der Warenzeichen- und Markenschutzgesetzgebung als frei zu betrachten wären und daher von jedermann benutzt werden dürften.

Verlag: Südwestdeutscher Verlag für Hochschulschriften Aktiengesellschaft & Co. KG
Dudweiler Landstr. 99, 66123 Saarbrücken, Deutschland
Telefon +49 681 37 20 271-1, Telefax +49 681 37 20 271-0
Email: info@svh-verlag.de
Zugl.: Berlin, TU, Diss., 2010

Herstellung in Deutschland:
Schaltungsdienst Lange o.H.G., Berlin
Books on Demand GmbH, Norderstedt
Reha GmbH, Saarbrücken
Amazon Distribution GmbH, Leipzig
ISBN: 978-3-8381-1745-4

Imprint (only for USA, GB)
Bibliographic information published by the Deutsche Nationalbibliothek: The Deutsche Nationalbibliothek lists this publication in the Deutsche Nationalbibliografie; detailed bibliographic data are available in the Internet at http://dnb.d-nb.de.

Any brand names and product names mentioned in this book are subject to trademark, brand or patent protection and are trademarks or registered trademarks of their respective holders. The use of brand names, product names, common names, trade names, product descriptions etc. even without a particular marking in this works is in no way to be construed to mean that such names may be regarded as unrestricted in respect of trademark and brand protection legislation and could thus be used by anyone.

Publisher: Südwestdeutscher Verlag für Hochschulschriften Aktiengesellschaft & Co. KG
Dudweiler Landstr. 99, 66123 Saarbrücken, Germany
Phone +49 681 37 20 271-1, Fax +49 681 37 20 271-0
Email: info@svh-verlag.de

Printed in the U.S.A.
Printed in the U.K. by (see last page)
ISBN: 978-3-8381-1745-4

Copyright © 2010 by the author and Südwestdeutscher Verlag für Hochschulschriften Aktiengesellschaft & Co. KG and licensors
All rights reserved. Saarbrücken 2010

Abstract

The present work treats several asymptotic properties of stochastic flows on $\mathbb{R}^d$, whose distributions are frequently assumed to be invariant under rotations of the state space. These stochastic flows include the isotropic Brownian flows, which have been studied since Yaglom [60] or Baxendale, Harris [5] and Le Jan [28]. Furthermore isotropic Ornstein-Uhlenbeck Flows are treated, which are considered since Dimitroff [12] as well as repulsive isotropic flows, which are about to be introduced in this work. All these classes of stochastic flows are linked by a single stochastic differential equation which passes to one of the named cases by the specification of one real parameter. First we define the classes of models and cite important facts from the literature.

Afterwards the spatial asymptotic behaviour of isotropic stochastic flows is treated in a short warm up chapter. A lemma is proved that serves to show that the unit ball based random dynamical system coming from an isotropic Ornstein-Uhlenbeck flow is not sufficiently smooth to apply well known results concerning Pesin's formula from Ledrappier and Young [32], [33] and hence to motivate the self contained study of this subject.

Another short chapter is concerned with the following question. When does the finite-point motion of a given stochastic flow admit a continuous (or even smooth) density which is strictly positive apart from the generalized diagonal? It is also shown that a large subclass of the isotropic flows belongs to the scope of the obtained results.

The following first main chapter treats the asymptotic behaviour of the shape of the set of points in $\mathbb{R}^d$ that has been visited up to some time. It is shown in the case of a planar isotropic Brownian flow that this shape is deterministic in probability. Dolgopyat, Kaloshin and Koralov give a similar result in a different setting ([13]). But since the core of their proof - the spatial periodicity of their model - fails to hold for all isotropic flows it has to be replaced by a different feature of the isotropic Brownian flows namely their invariance properties w.r.t. time reversion.

The so-called Margulis-Ruelle inequality asserts that the entropy of a random dynamical system can be estimated from above through the sum of its positive Lyapunov exponents. This inequality is extended to the case of a random dynamical system coming from an isotropic Ornstein-Uhlenbeck flow.

The last two main chapters are devoted to the asymptotic expansion of the spatial derivative of a stochastic flow taking the supremum over a compact set of initial points in $\mathbb{R}^d$. It is shown that this expansion is at most exponentially fast in time and a deterministic bound on the expansion speed is obtained. This result can be seen as a first step towards Pesin's formula for isotropic Ornstein-Uhlenbeck flows. First the case of first order derivatives of an isotropic flow is treated and afterwards the result is generalized - with worse constants - to a much more general class of stochastic flows and derivatives of arbitrary order.

Finally some open questions are listed and possible directions of further research are discussed.

Acknowledgement

I thank Prof. Dr. Michael Scheutzow for supervising this thesis, Prof. Dr. Peter Friz for accepting to be a co-examiner and Prof. Dr. Rolf Möhring for accepting to chair the PhD. examination. I also thank my colleagues (and former colleagues) especially Dr. Georgi Dimitroff and Moritz Biskamp from the stochastics group at TU Berlin for useful discussions and suggestions during the years. Furthermore I thank Anika Frischwasser for reading the manuscript and for the resulting improvement in orthography and transparency. Partial financial support from the International Research Training Group „Stochastic Models Of Complex Processes" is gratefully acknowledged.

Berlin, June 23, 2010
Holger van Bargen

Contents

		Introduction And Main Results	1
1		**Definition Of The Models**	**7**
	1.1	Stochastic Flows And Stochastic Differential Equations	7
		1.1.1 The Driving Fields And Their Characteristic	8
		1.1.2 Kunita-Type Integrals And Stochastic Flows	12
	1.2	Stochastic Flows As Random Dynamical Systems	18
		1.2.1 Definitions Of The Basic Notions About RDS'	19
		1.2.2 Construction Of A RDS From A Stochastic Flow	23
		1.2.3 Time Discretization, Markov Chain Representations	25
	1.3	Isotropic Brownian Flows	26
		1.3.1 Covariance Tensors And Correlation Functions	27
		1.3.2 Brownian Fields And Generated Flows	29
		1.3.3 Lyapunov Exponents	34
		1.3.4 Gaussian Measures And Reproducing Kernel Hilbert Spaces	35
		1.3.5 Time Reverse And Markov-Properties	38
	1.4	Isotropic Ornstein-Uhlenbeck Flows	38
		1.4.1 The Generating Field	40
		1.4.2 Invariant Measures	42
		1.4.3 Lyapunov Exponents	43
	1.5	Repulsive Isotropic Flows	44
2		**Spatial Regularity**	**47**
	2.1	Preliminaries	47
	2.2	Spatial Regularity Lemma	48

3 Densities For The Finite-Point Motions — 53
- 3.1 Existence Of Densities — 54
- 3.2 A-Little-Positivity Of Densities — 55
- 3.3 Positivity Of Densities — 57

4 A Weak Limit Shape Theorem For Planar Isotropic Brownian Flows — 61
- 4.1 Introduction And Preliminaries — 61
  - 4.1.1 Chasing Ball Property, LDP For Discrete Supermartingales — 62
  - 4.1.2 Sub-Gaussian Tails And Sublinear Growth — 64
- 4.2 Statement Of The Main Results — 64
- 4.3 The Lower Bound — 65
  - 4.3.1 Hitting Time Of Far Away Balls — 65
  - 4.3.2 Linear Expansion And Stable Norm — 81
  - 4.3.3 Sweeping Lemma And A Sharp Lower Bound - The Two-Dimensional Case — 85
  - 4.3.4 Dependence Of $\|v\|^R$ On R — 93
- 4.4 The Upper Bound — 95
  - 4.4.1 The Speed Of A Slow Curve Asymptotically Has A Dirac Distribution On The Proper Time Scale — 95
  - 4.4.2 Time Reverse - Comparison Of Fast And Slow Curves — 96

5 Dreaming Of Pesin's Formula — 103
- 5.1 Introduction — 103
- 5.2 Throwing An IOUF On A Ball - Stereographic Projection — 104

6 The Margulis-Ruelle Inequality for IOUFs — 109
- 6.1 Introduction — 109
- 6.2 Basics From Entropy Theory — 110
  - 6.2.1 The Definition Of The Metric Entropy Of An IOUF — 110
- 6.3 The Margulis-Ruelle Inequality — 115
  - 6.3.1 Statement Of The Result — 115
  - 6.3.2 Construction Of The Partitions — 116
  - 6.3.3 Entropy Estimates — 117
- 6.4 Exterior Powers, Tensor Algebras And SVDs — 121

7	**Asymptotic Growth Of Spatial Derivatives Of Isotropic Flows**	**123**
	7.1 Introduction And Preliminaries	123
	7.2 The Main Result	126
	7.3 Proof Of Lemma 7.2.2: The One-Point Condition	128
	7.4 Proof Of Lemma 7.2.3: The Two-Point Condition	134
	7.4.1 General Estimates And Preparation	134
	7.4.2 Derivation Of Formula H	139
	7.4.3 Treating Small $|x-y|$ And Large T	148
	7.4.4 Evaluation Of Formula H	153
	7.5 A New Proof For Linear Expansion	162
8	**Asymptotic Growth Of Spatial Derivatives: General Case**	**165**
	8.1 The Main Result	165
	8.2 Proof Of Lemma 8.1.2: The One-Point Condition	167
	8.3 Proof Of Lemma 8.1.3: The Two-Point Condition	170
	8.3.1 One-Point Mean Estimates	170
	8.3.2 Two-Point Estimates	170
9	**Further Steps To Pesin's Formula**	**175**
	9.1 Local Stable Manifolds	175
	9.2 Global Stable Manifolds	180
	9.3 Local Hölder Continuity Of Oseledec' Splittings	181
10	**Some Open Problems**	**185**
	10.1 Open Problems arising from Chapter 2	185
	10.2 Open Problems arising from Chapter 3	186
	10.3 Open Problems arising from Chapter 4	186
	10.4 Open Problems arising from Chapter 5	187
	10.5 Open Problems arising from Chapter 6 and Chapter 9	187
	10.6 Open Problems arising from Chapter 7 and Chapter 8	188
	Bibliography	**189**
	Index Of Notation and Abbreviations	**195**

Introduction And Main Results

The name of this work is „Some Asymptotic Properties Of Stochastic Flows". A stochastic flow is a random variable taking values in the groups of diffeomorphisms of the Euclidean space $\mathbb{R}^d$. It can be motivated by the following physical considerations. Suppose we have an infinite ocean without any prefered streams. This means that we suppose the local movement in this ocean to be equally likely to occur in any direction. On this ocean we now consider a set of specially distinguished points. One may think of an oil spill floating on the surface. This leads to the model of a so-called isotropic Brownian flow (IBF) which is specified by the invariance of its law unter rigid motions of the state space. These flows can be rigorously defined by a stochastic differential equation (SDE) of Kunita-type. These generically include an infinite-dimensional stochastic noise and generalize the notion of Itô SDEs. Unfortunately the full distribution of say the shape of the oil spill at some time T is too complex to be fully amenable for a detailed study. Hence we will have to restrict ourselves to the study of some macroscopic properties.

More generally we will not only consider the case of a zero drift environment i.e. of an isotropic ocean but also allow the motion to evolve in a quadratic potential. This potential may have a global minimum at the origin leading to the model of an isotropic Ornstein-Uhlenbeck flow (IOUF) or a maximum which yields the notion of a repulsive isotropic flow (RIF). These three classes of stochastic flows IBFs, IOUFs and RIFs and their properties are the main subject of the entire work. They are linked by the following SDE.

$$\phi_{s,t}(x) = x + \int_s^t M(du, \phi_{s,u}(x)) - c\int_s^t \phi_{s,u}(x)du.$$

Therein $M = M(t, x)$ is an isotropic Brownian field and describes the infinitesimal random kick that a particle located at $x \in \mathbb{R}^d$ experiences at time $t \geq 0$. The specification of $c \in \mathbb{R}$ leads to the different notions of stochastic flow which we already mentioned. The positive values for c yield the IOUFs, $c = 0$ corresponds to the IBFs and the $c < 0$ lead to the RIFs. In all cases $\phi_{s,t}(x)$ denotes the position of the particle at time t which was at position x at time s.

Chapter 1 is devoted to the precise definition of all the models and to the collection of important results from the literature. We also further develop some of the infrastructure results which are needed in the later chapters.

Afterwards Chapter 2 treats the question, how $\frac{\phi_{0t}(x) - \mathbb{E}[\phi_{0t}(x)]}{x}$ is behaved for large $|x|$. To be precise we show that this quantity converges to zero as $R \to \infty$ if we take the supremum over $R \leq |x| \leq R+1$. Although the scaling by x seems to be too large we stick to it because this is precisely the scaling we will need in Chapter 5. The convergence is proved by using the well-known chaining technique which originates from [10] and which we explain in some detail.

Chapter 3 has a scope that is somewhat different from the one of the rest of the work. We analyze the question when the solution of a SDE admits a density that is strictly positive on some predescribed set. To be precise we consider the SDEs for the finite-point motions and ask for the necessary amount of differentiability of the driving vector fields that guarantees the existence of a density. Afterwards we apply a criterion from [29] that allows us to show that the density (provided it exists) is strictly positive apart from the generalized diagonal i.e. the set of points in $(\mathbb{R}^d)^n$ where no two of the n elements coincide. We also give a notion of positivity that can be obtained with very elementary methods from isotropy properties of the flows we consider. Nevertheless this notion is not sufficient for the applications we have in mind and so we show that a large class of the isotropic flows considered here fall into the range of well-known results concerning strict positivity.

The first main Chapter 4 deals with the diameter problem in the case of a two-dimensional IBF Φ with a positive Lyapunov exponent. We show that there is a deterministic set $\mathcal{B}$ such that we have for any deterministic bounded initial set $\gamma \subset \mathbb{R}^2$ that consists of at least two different points the following. If we put $\gamma_t := \Phi_t(\gamma)$ and $\mathcal{W}_t(\gamma) := \bigcup_{0 \leq s \leq t} \gamma_s$ we have for any $\epsilon > 0$ that

$$\lim_{T\to\infty} \mathbb{P}\left[(1-\epsilon)T\mathcal{B} \subset \mathcal{W}_T(\gamma) \subset (1+\epsilon)T\mathcal{B}\right] = 1.$$

It is also shown that the lower bound holds a.s. for sufficiently large T. The chapter is divided into the proof of the lower bound and the proof of the upper one. The first part i.e. the lower bound is shown with a combination of methods from [13] and [53]. We reprove the fact that the diameter of a bounded set under the action of an IBF with a positive exponent grows linearly in time and define the asymptotic expansion speed in terms of a stable norm. Afterwards it is shown that a curve that reaches the R-neighbourhood of $P \in \mathbb{R}^2$ will sweep this neighbourhood within a rather short time with high probability.

The upper bound i.e. the fact that the lower expansion speed and the upper expansion speed coincide relies on the fact, that the forward distribution of an IBF and its backward distribution (i.e. the image measure obtained by time reversion) coincide. This fact is shown to imply the following. Suppose one has two curves with diameter say 17 which have distance say 42 from each other. Let both of the curves evolve under the action of two independent IBFs for some time $T > 0$. Then the probability that the curves intersect each other after time T can be bounded from below by a strictly positive constant p. This argument is also the main issue that necessitates the restriction to $d = 2$. This fact surely cannot be expected to be easily proved in the general d-dimensional case. This is also the point where the techniques from [13] turn out to be inappropriate to cover the case of IBFs.

The motivating Chapter 5 shows that the random dynamical system (RDS) that can be obtained by an IOUF via stereographic projection on the unit ball fails to be continuously differentiable. Nevertheless this is necessary to be able to apply the standard results from the literature obtained by Ledrappier and Young in their papers [32], [33] and [34]. The explicit computations performed in Chapter 5 show that it is completely hopeless to expect that these results in their standard versions apply to the RDSs coming from an IOUF. Here the spatial regularity lemma from Chaper 2 is used in the central step.

Hence we have to work on these sophisticated results on our own replacing the compactness assumption of Ledrappier and Young by different features of our models. The first part of Pesin's formula asserting that the entropy of a $C^{1+\epsilon}$-RDS on a compact manifold equals the sum of its positive Lyapunov exponents counted with their multiplicities is the subject of Chapter 6. Therein we prove that the "≤"-part of this equality - also called Margulis-Ruelle inequality in the literature - is true in the case of an IOUF. We prove this via the combination of the method from Bahnmüller and Bogenschütz [3] with an exhaustion argument for the construction of a suitable partition in the entropy estimation. One might argue that an inequality is not an asymptotic statement but the Margulis-Ruelle inequality is an inequality between two asymptotic quantities. The entropy links to the asymptotic number of guesses one needs to find the path of a particle given the randomness and the Lyapunov exponents give the asymptotic exponential growth rate of the singular values of the spatial derivative at a previously fixed point.

The main Chapter 7 treats the asymptotic growth of the spatial derivatives of an isotropic stochastic flow. We show that for a compact $\Xi \subset \mathbb{R}^d$ we have for $\psi_t(x) := \log ||D\phi_t(x)||$ that
$$\limsup_{T \to \infty} \left(\sup_{0 \leq t \leq T} \sup_{x \in \Xi} \frac{1}{T} |\psi_t(x)| \right) \leq K \text{ a.s.}$$
where K depends on the box dimension of Ξ and on d. The proof of this fact relies on the chaining growth theorem obtained by Scheutzow in [52] which was taylored to the case of the flow itself but proves to be sufficient to cover the case of the spatial derivatives of an isotropic flow. This theorem requires to have bounds on the growth of the one-point motion and to have bounds on the two-point distances. While the first part i.e. the one point estimate can be obtained with a straightforward computation the two-point estimate turns out to be very technical. It also involves fourth order Taylor expansions of the covariance tensor b in the present form. This fact seems to prevent the result from being easily generalized to a wider class of stochastic flows with the applied method.

Chapter 8 treats the same problem as Chapter 7 in much more general setting. It extends the result (exponential growth of derivatives) not only to a much more general class of stochastic flows but also proves it for derivatives of arbitrary order. The proof relies on estimates from Imkeller and Scheutzow [20] and is much shorter than the one given in Chapter 7. Nevertheless the result is less explicit concerning the constants appearing and it does not cover the approach of first-order derivatives to singularity because it uses $\psi_t(x) := \log(1 + ||D^i\phi_t(x)||)$.

The outline in Chapter 9 gives some more steps towards Pesin's formula for IOUFs along the lines of Liu and Qian [37]. The local Hölder continuity of the Oseledec' splittings is shown.

In the final Chapter 10 we list some open problems and give some conjectures on their solution. It indicates directions for further research.

Chapter 1

Definition Of The Models

In this chapter we introduce the general notion of a stochastic flow and specify the conditions leading to the models we treat in the sequel. This includes the notions of isotropic covariance tensors, spatial semimartingales and stochastic differential equations driven by them. We also review important infrastructural results e.g. diffuse behaviour of finite-dimensional projections, support theorems and the expansion in terms of the reproducing kernel Hilbert spaces. The chapter is divided into several sections. First we give a brief review on so called Kunita-integration and the link between stochastic flows and stochastic differential equations (SDEs) of Kunita-type. Then we explain how to obtain a random dynamical system (RDS) from a given stochastic flow. In the later sections we are more specific to provide information on the particular models we use. We give the definitions of isotropic Brownian flows (IBFs), isotropic Ornstein-Uhlenbeck flows (IOUFs) and repulsive isotropic flows (RIFs) and note some of their important properties.

1.1 Stochastic Flows And Stochastic Differential Equations

All definitions and statements in this section originate from [27]. We will essentially follow [12] in the first part of the exposition. As usual we always assume that we have an appropriate probability space $(\Omega, \mathcal{F}, \mathbb{P})$ on which all the random variables are defined.

1.1.1 The Driving Fields And Their Characteristic

Let $d, n \in \mathbb{N}$. For a domain $\mathbb{D} \subset \mathbb{R}^d$ and a non-negative integer m denote by $C^m(\mathbb{D} : \mathbb{R}^n)$ the set of all functions $f \colon \mathbb{D} \to \mathbb{R}^n$ which are m-times continuously differentiable. If $m = 0$ then the superscript may be omitted. As usual for a multi-index of non-negative integers $\alpha := (\alpha_1, \ldots, \alpha_n)$ we put $|\alpha| = \alpha_1 + \cdots + \alpha_n$ and define the differential operator

$$\mathrm{D}_x^\alpha := \frac{\partial^{|\alpha|}}{(\partial x_1)^{\alpha_1} \ldots (\partial x_n)^{\alpha_n}}.$$

Let $\mathbb{K} \subset \mathbb{D}$ be a compact set. Define the seminorms

$$\|f\|_{m:\mathbb{K}} := \sup_{x \in \mathbb{K}} \frac{|f(x)|}{1 + |x|} + \sum_{1 \leq |\alpha| \leq m} \sup_{x \in \mathbb{K}} |\mathrm{D}_x^\alpha f(x)|,$$

where $|\cdot|$ denotes the Euclidean distance in the corresponding space. We denote $\|\cdot\|_{m:\mathbb{D}}$ by $\|\cdot\|_m$.

The family of seminorms $\{\|\cdot\|_{m:\mathbb{K}} : \mathbb{K} \subset \mathbb{D}$ is compact$\}$ turns $C^m(\mathbb{D} : \mathbb{R}^n)$ into a Fréchet space (metrizable locally convex topological vector space or equivalently locally convex topological vector space, whose topology is generated by countably many seminorms). Further the set

$$C_b^m(\mathbb{D} : \mathbb{R}^n) := \{f \in C^m(\mathbb{D} : \mathbb{R}^n) : \|f\|_m < \infty\}$$

equipped with the norm $\|\cdot\|_m$ is a Banach space.

For arbitrary $\delta \in [0,1]$ let $C^{m,\delta}(\mathbb{D} : \mathbb{R}^n)$ be the set of all $f \in C^m(\mathbb{D} : \mathbb{R}^n)$ such that all $\mathrm{D}_x^\alpha f$ for $|\alpha| = m$ are Hölder continuous with exponent $\delta \geq 0$. The set $C^{m,\delta}(\mathbb{D} : \mathbb{R}^n)$ with the seminorms

$$\|f\|_{m+\delta:\mathbb{K}} := \|f\|_{m:\mathbb{K}} + \sum_{|\alpha|=m} \sup_{y,z \in \mathbb{K},\, z \neq y} \frac{|\mathrm{D}_x^\alpha f(y) - \mathrm{D}_x^\alpha f(z)|}{|y-z|^\delta}$$

is again a Fréchet space. $\|\cdot\|_{m+\delta:\mathbb{D}}$ will be denoted by $\|\cdot\|_{m+\delta}$.

A continuous function $f(x,t) \colon \mathbb{D} \times \mathbb{R}_+ \to \mathbb{R}^n$ is said to belong to the class $C^{m,\delta}$ if

1. $f(t) := f(\cdot, t) \in C^{m,\delta}$ for all $t \in \mathbb{R}_+$ and

1.1 Stochastic Flows And Stochastic Differential Equations

2. $\int_0^T \|f(t)\|_{m+\delta:\mathbb{K}}\, dt < \infty$ for all compact $\mathbb{K} \subset \mathbb{D}$ and all $T < \infty$.

Further $f(x,t)$ is said to belong to the class $C_b^{m,\delta}$ if 2. is replaced by

2.' $\int_0^T \|f(t)\|_{m+\delta:\mathbb{D}}\, dt = \int_0^T \|f(t)\|_{m+\delta}\, dt < \infty$ for all $T < \infty$.

and $f(x,t)$ is said to belong to the class $C_{ub}^{m,\delta}$ if 2. is replaced by

2." $\sup_{t \in \mathbb{R}_+} \|f(t)\|_{m+\delta:\mathbb{D}} = \sup_{t \in \mathbb{R}_+} \|f(t)\|_{m+\delta} < \infty$.

The notations C^m, C_b^m and C_{ub}^m are abbreviations for $C^{m,0}$, $C_b^{m,0}$ and $C_{ub}^{m,0}$ respectively. Now we consider functions of the type $g \colon \mathbb{D} \times \mathbb{D} \to \mathbb{R}^n$. Let $\tilde{C}^m$ be the set of all functions g, which are m-times continuously differentiable with respect to each variable x and y. For $g \in \tilde{C}^m$ define the seminorms

$$\|g\|_{m:\mathbb{K}}^{\sim} := \sup_{x,y \in \mathbb{K}} \frac{|g(x,y)|}{(1+|x|)(1+|y|)} + \sum_{1 \leq |\alpha| \leq m} \sup_{x,y \in \mathbb{K}} |D_y^\alpha D_x^\alpha g(x,y)|$$

and for $\delta \in (0,1)$

$$\|g\|_{m+\delta:\mathbb{K}}^{\sim} := \|g\|_{m:\mathbb{K}}^{\sim} + \sum_{|\alpha|=m} \|D_x^\alpha D_y^\alpha g\|_{\delta:\mathbb{K}}^{\sim},$$

where

$$\|g\|_{\delta:\mathbb{K}}^{\sim} := \sup \left\{ \frac{|g(x,y) - g(x',y) - g(x,y') + g(x',y')|}{|x-x'|^\delta |y-y'|^\delta} : x, y, x', y' \in \mathbb{K},\ x \neq x',\ y \neq y' \right\}.$$

Again if $\mathbb{K} = \mathbb{D}$ then it is omitted from the subscript in the notation of the seminorms. We set

$$\tilde{C}_b^m := \{g \in C(\mathbb{D} \times \mathbb{D} : \mathbb{R}^n) : \|g\|_m^{\sim} < \infty\} \quad \text{and} \quad \tilde{C}_b^{m,\delta} := \{g \colon \mathbb{D} \times \mathbb{D} \to \mathbb{R}^n : \|g\|_{m+\delta}^{\sim} < \infty\}.$$

A continuous function $g \colon \mathbb{D} \times \mathbb{D} \times [0,T] \to \mathbb{R}^n$ is said to belong to the class $\tilde{C}^{m,\delta}$ if

1. $g(t) := g(\cdot, \cdot, t) \in \tilde{C}^{m,\delta}$ for all $t \in [0,T]$ and

2. $\int_0^T \|g(t)\|_{m+\delta:\mathbb{K}}^{\sim}\, dt < \infty$ for all compact $\mathbb{K} \subset \mathbb{D}$ and $0 < T < \infty$.

The classes $\tilde{C}_b^{m,\delta}$ and $\tilde{C}_{ub}^{m,\delta}$ are defined by replacing 2. in the same manner as above (of course with $||.||^\sim$ instead of $||.||$.). The definitions of the classes $\tilde{C}^m$, $\tilde{C}_b^m$ and $\tilde{C}_{ub}^m$ are also obtained by putting $\delta = 0$ as before . Let now $(\Omega, (\mathcal{F}_t)_{t\geq 0}, \mathcal{F}, \mathbb{P})$ be a filtered probability space satisfying the usual conditions (i.e. right-continuity and completeness). Consider a random field

$$F(t, x, \omega) \colon \mathbb{R}_+ \times \mathbb{D} \times \Omega \to \mathbb{R}^n.$$

If F is m-times continuously differentiable with respect to x for any t almost surely, we can regard it as a C^m-valued stochastic process. If it is continuous we call it a continuous C^m-valued stochastic process. Analogously we can define $C^{m,\delta}$-valued and continuous $C^{m,\delta}$-valued stochastic processes.

Let now

$$G(t, x, y, \omega) \colon [0, T] \times \mathbb{D} \times \mathbb{D} \times \Omega \to \mathbb{R}^n.$$

Similarly as above we say G is a $\tilde{C}^m$-valued stochastic process if G is m-times continuously differentiable with respect to both x and y almost surely for all $t \in [0, T]$. Analogously we can define continuous $\tilde{C}^m$-valued, $\tilde{C}^{m,\delta}$-valued and continuous $\tilde{C}^{m,\delta}$-valued processes.

Assume now that $M(t, x, \omega) \colon \mathbb{R}_+ \times \mathbb{D} \times \Omega \to \mathbb{R}^d$ is a family of continuous local martingales, indexed by $x \in \mathbb{D}$ with $M(0, x) \equiv 0$ a.s.. The next theorem gives the connection between the regularity with respect to the spatial variable of M and the one of $\langle M(\cdot, x), M(\cdot, y)\rangle_t$.

Theorem 1.1.1 (continuous modifications).

1. Let $M(t, x)$, $x \in \mathbb{D}$ be a family of continuous local martingales with $M(0, x) \equiv 0$ $\mathbb{P}$-a.s.. If the joint quadratic variation $\langle M(\cdot, x), M(\cdot, y)\rangle_t$ has a modification of a continuous $\tilde{C}^{m,\delta}$-process for some $m \geq 1$ and $\delta \in (0, 1]$ then $M(x, t)$ has a modification of a continuous $C^{m,\epsilon}$-process for any $\epsilon < \delta$. Furthermore, for a multi-index α with $|\alpha| \leq m$, $D_x^\alpha M(t, x)$ is a family of continuous local martingales with joint quadratic variation process $D_x^\alpha D_y^\alpha \langle M(\cdot, x), M(\cdot, y)\rangle_t$. Such a family of continuous local martingales is called a continuous $C^{m,\epsilon}$-local martingale.

2. Let $M(t, x)$, $x \in \mathbb{D}$ and $N(t, y)$, $y \in \mathbb{D}$ be continuous local martingales with values in $C^{m,\delta}$, where $m \geq 0$ and $\delta > 0$. Then the joint quadratic variation $\langle M(\cdot, x), N(\cdot, y)\rangle_t$ has a modification of a continuous $\tilde{C}^{m,\epsilon}$-process for any $\epsilon < \delta$. Moreover, the modi-

1.1 Stochastic Flows and Stochastic Differential Equations

fication satisfies

$$D_x^\alpha D_y^\beta \langle M(\cdot, x), N(\cdot, y) \rangle_t = \langle D_x^\alpha M(\cdot, x), D_y^\beta N(\cdot, y) \rangle_t.$$

Proof: [27, Theorem 3.1.2 and Theorem 3.1.3]. □

Let $F(t, x)$, $x \in \mathbb{D}$ be a family of continuous semimartingales decomposed as $F(t, x) = M(t, x) + V(t, x)$ where $M(t, x)$ is a continuous local martingale for any $x \in \mathbb{R}^d$ and $V(t, x)$ is a continuous process of bounded variation for any $x \in \mathbb{R}^d$. A family of continuous semimartingales $F(t, x)$ is called a $C^{m,\delta}$-semimartingale if $M(t, x)$ is a $C^{m,\delta}$-local martingale and $V(t, x)$ is a continuous $C^{m,\delta}$-process, such that for any $x \in \mathbb{R}^d$ and $|\alpha| \leq m$ the process $t \mapsto D_x^\alpha V(t, x)$, is of bounded variation. Assume that

$$\langle M(\cdot, x), M(\cdot, y) \rangle_t = \int_0^t a(x, y, s) ds \quad \text{and} \quad V(t, x) = \int_0^t b(x, s) ds$$

holds for functions $a(x, y, s) = (a^{ij}(x, y, s))_{1 \leq i, j \leq n}$ and $b(x, s) = b^i(x, s)_{1 \leq i \leq n}$. The pair of random fields (a, b) is called the local characteristic of the semimartingale field $F(t, x)$. [27] allows for a and b to be densities w.r.t. a different measure but Lebesgue measure, but we do not make use of that. The pair (a, b) is said to belong to the class $(B^{m,\delta}, B^{m',\delta'})$ if

1. $a(x, y, t)$ has a modification that is a predictable process with values in $\tilde{C}^{m,\delta}$.
2. $\int_0^T \|a(x, y, t)\|_{m+\delta:\mathbb{K}}^\sim dt < \infty$ P-almost surely for all compact $\mathbb{K} \subset \mathbb{D}$ and all $T < \infty$.
3. $b(x, t)$ has a modification that is a predictable process with values in $C^{m',\delta'}(\mathbb{D} : \mathbb{R})$.
4. $\int_0^T \|b(x, t)\|_{m'+\delta':\mathbb{K}} dt < \infty$ P-almost surely for all compact $\mathbb{K} \subset \mathbb{D}$ and all $T < \infty$.

Further, the pair (a, b) is said to belong to class $(B_b^{m,\delta}, B_b^{m',\delta'})$ if 2. and 4. are replaced by

2.' $\int_0^T a(t) dt := \int_0^T \|a(x, y, t)\|_{m+\delta:\mathbb{D}}^\sim dt < \infty$ P-almost surely for all $T < \infty$.

4.' $\int_0^T b(t) dt := \int_0^T \|b(x, t)\|_{m'+\delta':\mathbb{D}} dt < \infty$ P-almost surely for all $T < \infty$.

(a, b) is said to belong to the class $(B_{ub}^{m,\delta}, B_{ub}^{m',\delta'})$ if 2. and 4. are replaced by the condition that $\|a(x, y, t)\|_{m+\delta:\mathbb{D}}^\sim$ and $\|b(x, t)\|_{m'+\delta':\mathbb{D}}$ are uniformly bounded for all $t \in \mathbb{R}^+$ and $\omega \in \Omega$. If $m = m'$ and $\delta = \delta'$ we say that (a, b) belongs to the class $B^{m,\delta}$ (respectively $B_b^{m,\delta}$ and $B_{ub}^{m,\delta}$).

A continuous $C^{m,\delta}$-valued process is called a $C^{m,\delta}$-*valued Brownian motion* if for any partition $0 \leq t_0 \leq t_1 \leq \cdots \leq t_n$ the family $(F(t_{i+1}, \cdot) - F(t_i, \cdot))_{i=0,\ldots,n-1}$ of random variables with values in $C^{m,\delta}$ is independent.

1.1.2 Kunita-Type Integrals And Stochastic Flows

For a domain $\mathbb{D} \subset \mathbb{R}^d$, let $F(t, x, \omega)\colon \mathbb{R}_+ \times \mathbb{D} \times \Omega \to \mathbb{R}^n$ be a family of continuous real valued semimartingales decomposed as $F(t,x) = M(t,x) + V(t,x)$ where $M(t,x)$ is a $C^{m,\delta}$-local martingale and $V(t,x)$ is a continuous $C^{m,\delta}$-process, such that $D_x^\alpha V(t,x)$, $|\alpha| \leq m$ are all processes of bounded variation. Assume again that the local characteristic (a,b) can be written as

$$\langle M(\cdot, x), M(\cdot, y) \rangle_t = \int_0^t a(x,y,s)ds \quad \text{and} \quad V(t,x) = \int_0^t b(x,s)ds$$

(with $a(x,y,s)$ and $b(x,s)$ as before) and belongs to the class $B^{0,\delta}$ for some $\delta > 0$. Let $\{f_t : t \geq 0\}$ be a predictable process with values in $\mathbb{D}$, satisfying

$$\int_0^T a(f_s, f_s, s)ds < \infty \quad \text{and} \quad \int_0^T |b(f_s, s)|ds < \infty \quad \text{P-a.s. for all } T < \infty. \tag{1.1}$$

The stochastic integral of f based on the semimartingale kernel F is defined for arbitrary $t \geq 0$ by $\int_0^t F(ds, f_s) = \int_0^t M(ds, f_s) + \int_0^t b(f_s, s)ds$, where the integral $\int_0^t M(ds, f_s)$ (and its Kunita-Stratonovich or Kunita-Itô backward counterparts) is defined similarly to the well known classical stochastic integrals in two steps.

1st Step: For simple predictable processes, i.e. processes of the form

$$f_t = \sum_{k=0}^n f_{t_k} 1\!\!1_{(t_k, t_{k+1}]}(t) + f_0 1\!\!1_{\{t=0\}},$$

1.1 Stochastic Flows And Stochastic Differential Equations

satisfying (1.1) the stochastic integral of f based on the kernel M is defined via

$$\int_0^t M(ds, f_s) := \sum_{k=0}^n M(t_{k+1} \wedge t, f_{t_k \wedge t}) - M(t_k \wedge t, f_{t_k \wedge t}) \text{ (Kunita-Itô version)},$$

$$\int_0^t M(\circ ds, f_s) := \frac{1}{2} \sum_{k=0}^n M(t_{k+1} \wedge t, f_{t_{k+1} \wedge t}) + M(t_{k+1} \wedge t, f_{t_k \wedge t}) - M(t_k \wedge t, f_{t_{k+1} \wedge t})$$
$$- M(t_k \wedge t, f_{t_k \wedge t}) \text{ (Kunita-Stratonovich version)},$$

$$\int_0^t M(\hat{d}s, f_s) := \sum_{k=0}^n M(t_{k+1} \vee 0, f_{t_{k+1} \vee 0}) - M(t_k \vee 0, f_{t_{k+1} \vee 0}) \text{ (Kunita-Itô backward version)}.$$

2nd Step: For general predictable processes f_s satisfying (1.1) one determines an approximating Cauchy sequence of simple processes f^n, such that

$$\int_0^T (a(f_s^n, f_s^n, s) - 2a(f_s^n, f_s^m, s) + a(f_s^m, f_s^m, s)) ds \xrightarrow[n,m \to \infty]{} 0.$$

Then it can be shown (see [27, page 82-83]) that the sequence $\left(\int_0^\cdot M(ds, f_s^n) \right)_n$ converges in probability, uniformly in t on compact subsets of $\mathbb{R}_+$. The limiting continuous local martingale is defined to be the stochastic integral $\int_0^t M(ds, f_s)$. We omit the second step for the other notions of stochastic integral indicated in the first step because it is perfectly similar to the Kunita-Itô case (with the difference that the limits are no longer necessarily local martingales). Note that one has to require the local characteristic to be of class $(B^{2,\delta}, B^{1,0})$ for some $\delta > 0$ to define the Kunita-Stratonovich version of stochastic integral. The connection between the Itô and the Stratonovich integrals is given by the following proposition.

Proposition 1.1.2 (Kunita-Itô and Kunita-Stratonovich integrals).
Assume that $F(t,x)$ is a continuous C^1-semimartingale with local characteristic belonging to the class $(B^{2,\delta}, B^{1,0})$ for some $\delta > 0$ and that $\{f_t : t \geq 0\}$ is a continuous semimartingale. Then the Stratonovich integral is well defined and is related to the Itô integral by

$$\int_0^t F(\circ ds, f_s) = \int_0^t F(ds, f_s) + \frac{1}{2} \sum_{j=1}^d \left\langle \int_0^\cdot \frac{\partial F}{\partial x^j}(ds, f_s), f_\cdot^j \right\rangle_t.$$

Proof: [27, Theorem 3.2.5] □

The choice of a certain type of stochastic integral describing a stochastic flow through an SDE is a matter of taste (if both of them are defined) because both - the Kunita-Itô formulation and the Kunita-Stratonovich formulation - have advantages and disadvantages: the Kunita-Itô version preserves the „local martingale-property" of the inegrator and the Kunita-Stratonovich-formulation yields better symmetry properties w.r.t. time reversion. A much more detailed description of the stochastic calculus involving these types of integrals can be found in [27]. Note that we will mostly use the Kunita-Itô version of Kunita-type stochastic integration. Let now $F(t,x,\omega) \colon \mathbb{R}_+ \times \mathbb{R}^d \times \Omega \to \mathbb{R}^d$ be a continuous $C(\mathbb{R}^d : \mathbb{R}^d)$-valued semimartingale. Having the notion of the Kunita integrals one can consider SDEs involving such integrals

$$\phi_{s,t}(x) = x + \int_s^t F(du, \phi_{s,u}(x)) \quad \text{for } t \geq s \geq 0, \ x \in \mathbb{R}^d, \tag{1.2}$$

which may be interpreted as the equations describing the movement of a passive tracer in the random vector field $F(t,x)$ starting at time s in the location x. With the help of the well known Picard-iteration technique one can prove as in the classical case existence and pathwise uniqueness results for solutions of (1.2).

Theorem 1.1.3 (existence of solution to SDEs).
Assume that $F(t,x)$ is a continuous semimartingale with values in $C(\mathbb{R}^d : \mathbb{R}^d)$ and local characteristic belonging to the class $B_b^{0,1}$. Then for each $x \in \mathbb{R}^d$ and $s \geq 0$ the SDE (1.2) has a pathwise unique solution.

Proof: [27, Theorem 3.4.1] □

We now give a short introduction to stochastic flows and how they are to be considered as random variables.

Definition 1.1.4 (stochastic flow, Brownian flow).
A mapping $\phi_{s,t}(x,\omega) \colon \mathbb{R}_+ \times \mathbb{R}_+ \times \mathbb{R}^d \times \Omega \to \mathbb{R}^d$ is called a stochastic flow of homeomorphisms if for $\mathbb{P}$-almost all $\omega \in \Omega$ the following is true.

1. *$\phi_{s,t}(\omega)$ is continuous w.r.t. s, t and x.*

2. *$\phi_{s,t}(\omega) = \phi_{u,t}(\omega) \circ \phi_{s,u}(\omega)$ holds for all $s,t,u \in \mathbb{R}_+$.*

3. *$\phi_{s,s}(\omega) = id_{\mathbb{R}^d}$ for all $s \in \mathbb{R}_+$.*

4. $\phi_{s,t}(\omega)\colon \mathbb{R}^d \to \mathbb{R}^d$ is a homeomorphism for all $s,t \in \mathbb{R}_+$.

If additionally ϕ satisfies also 4. it is called a stochastic flow of C^k-diffeomorphisms.

4. $\phi_{s,t}(\omega)$ is k-times continuously differentiable with respect to x for all $s,t \in \mathbb{R}_+$.

A stochastic flow of homeomorphisms is called Brownian *if it has independent increments, that is for all $n \in \mathbb{N}$ and $0 \leq t_1 < \cdots < t_n$ the family of random mappings $\left(\phi_{t_1,t_2},\ldots,\phi_{t_{n-1},t_n}\right)$ is independent.*

Notation: Omitting an argument in the notation of ϕ means that we refer to ϕ as a function with respect to this argument, i.e. $\phi_{s,t}$ denotes the random homeomorphism $\omega \mapsto \bigl(x \mapsto \phi_{s,t}(x,\omega)\bigr)$. However, omitting the first time parameter means it is set to 0, i.e. $\phi_t(x) := \phi_{0,t}(x)$. Let G be the set of all homeomorphisms on $\mathbb{R}^d$. It has a group structure with respect to the composition of maps. One can introduce a metric on G via

$$d(\phi,\psi) = \rho(\phi,\psi) + \rho(\phi^{-1},\psi^{-1}),$$

where

$$\rho(\phi,\psi) := \sum_{N=1}^{\infty} \frac{\sup_{|x|\leq N}|\phi(x)-\psi(x)|}{2^N(1+\sup_{|x|\leq N}|\phi(x)-\psi(x)|)}.$$

Then (G,d) is a complete separable topological group. A stochastic flow of homeomorphisms can be viewed as a continuous random field with index set $\mathbb{R}_+ \times \mathbb{R}_+$ and values in G. It is called a *continuous stochastic flow with values in G*.
Let $G^k \subset G$ be the subgroup of all C^k-diffeomorphisms. Define the metric

$$d_k(\phi,\psi) := \sum_{|\alpha|\leq k} \rho(\mathrm{D}^\alpha \phi, \mathrm{D}^\alpha \psi) + \sum_{|\alpha|\leq k} \rho(\mathrm{D}^\alpha \phi^{-1}, \mathrm{D}^\alpha \psi^{-1}).$$

Then (G^k,d_k) is again a complete separable topological group. A stochastic flow of C^k-diffeomorphisms can be viewed as a continuous random field with index set $\mathbb{R}_+ \times \mathbb{R}_+$ and values in G^k. It is called a *continuous stochastic flow with values in G^k* or sometimes a *continuous flow of C^k-diffeomorphisms*.

The main topic in [27] is the relation between stochastic flows and SDEs of the type (1.2). This relation is also the main motivation for considering stochastic integrals of Kunita-type.

Assuming some regularity conditions, there is a one to one correspondence between the forward stochastic semimartingale flows with values in G and the continuous semimartingales with values in C through SDEs of type (1.2). Especially there are examples of very natural stochastic flows, which cannot be generated via classical SDEs, driven by finitely many Brownian motions, e.g. isotropic Brownian and isotropic Ornstein-Uhlenbeck flows (which both are of particular interest for this work). This one to one correspondence is established in [27, Theorem 4.4.1 and Theorem 4.5.1]. Here we quote another theorem from the same source which maintains one of the directions in the correspondence and also yields the smoothness of the solution of the SDE depending on the smoothness of the driving semimartingale field.

Theorem 1.1.5 (flow property of solutions of SDEs, generator of a stochastic flow).
Assume that the local characteristic of the continuous C-semimartingale $F(t,x)$ belongs to the class $B_b^{k,\delta}$ for some $k \geq 1$ and $\delta > 0$. Then the solution of the SDE (1.2) has a modification $(\phi_{s,t}(x) : 0 \leq s \leq t, x \in \mathbb{D})$ which is a forward semimartingale stochastic flow of C^k-diffeomorphisms, or a stochastic flow of C^k-diffeomorphisms. Further it is a $C^{k,\epsilon}$-semimartingale for any $0 < \epsilon < \delta$. In this case F is called a forward Itô's random infinitesimal generator of the flow $(\phi_{s,t}(x) : 0 \leq s \leq t, x \in \mathbb{D})$.

Proof: [27, Theorem 4.6.5] □

Remark 1.1.6. *Actually, the original statement of the above theorem is for a finite time horizon only, i.e. for $0 \leq s \leq t \leq T$ for some $T < +\infty$. However a standard localizing argument immediately generalizes the statement to arbitrary $s \leq t \in \mathbb{R}_+$.*

Occasionally we will use the notion of a forward and a backward stochastic flow meaning the forward ($s \leq t$) or the backward ($s \geq t$) restriction of $\phi_{s,t} \colon \mathbb{R}_+ \times \mathbb{R}_+ \times \Omega \to G$. The relation between the forward flow and the backward flow and their generating fields respectively is given by the following proposition.

Proposition 1.1.7 (link between forward flows and backward flows - general result).
Let $(\phi_{s,t}, 0 \leq s \leq t \leq T)$ be a forward Brownian flow of homeomorphisms with values in G^k for $k \geq 1$ satisfying the following conditions.

1. For each $0 \leq s \leq t < \infty$ and $x, y \in \mathbb{R}^d$ the random variable $\phi_{s,t}(x)$ is square integrable and we have

$$\lim_{h \to 0} \frac{\mathbb{E}\left[\phi_{t,t+h}(x) - x\right]}{h} = b(x,t) \text{ and}$$

$$\lim_{h \to 0} \frac{\mathbb{E}\left[(\phi_{t,t+h}(x) - x)(\phi_{t,t+h}(y) - y)^t\right]}{h} = a(x,y,t)$$

wherein (a, b) is the local characteristic of a forward Itô's random infinitesimal generator of ϕ, again denoted by F.

2. There exists a positive constant K such that we have for any $0 \leq s \leq t \leq T$ and $x, y \in \mathbb{R}^d$ that

$$\mathbb{E}\left[\phi_{s,t}(x) - x\right] \leq K(1 + |x|)|t - s|,$$
$$\mathbb{E}\left[(\phi_{s,t}(x) - x)(\phi_{s,t}(y) - y)^t\right] \leq K(1 + |x|)(1 + |y|)|t - s|.$$

3. We have for the local characteristic (a, b) that $a(x, y, t) \in \tilde{C}_{ub}^{k,\delta}$ and $b(x, t) \in C_{ub}^{k,\delta}$ where $k \geq 2$ and $\delta > 0$.

Define the C^{k-1}-valued Brownian motion $\hat{F}(x,t) = F(x,t) - \int_0^t \sum_{j=1}^d \frac{\partial a^{\cdot j}}{\partial x_j}(x,y,s)|_{y=x} ds$. Then the associated backward flow $\phi_{t,s} \equiv (\phi_{s,t}^{-1}, 0 \leq s \leq t \leq T)$ is governed by a Kunita-Itô backward SDE based on $\hat{F}$, i.e. it satisfies for $x \in \mathbb{R}^d$ and $0 \leq s \leq t \leq T$ that

$$\phi_{t,s}(x) = x - \int_s^t \hat{F}(\phi_{t,r}(x), \hat{d}r).$$

Proof: [27, Theorem 4.2.10]. Observe that it is sufficient to require the limits in 1. to exist, because if they exist, they equal the local characteristic of F (this is the way a and b are originally defined in [27]). □

Note that the correction to be applied when passing from the Kunita-Itô forward formulation of the flow equation to the backward equation is just two times the correction that is to be applied when passing from Kunita-Itô forward formulation to Kunita-Stratonovich forward formulation. We end this section stating the Markov properties we have to use repeatedly in the following lemma.

Lemma 1.1.8 (Markov properties of stochastic flows).
Let $\phi_{s,t}(x,\omega)\colon \mathbb{R}_+ \times \mathbb{R}_+ \times \mathbb{R}^d \times \Omega \to \mathbb{R}^d$ be a forward Brownian flow of homeomorphisms (on $(\Omega, \mathcal{F}, \mathbb{P})$) and let $\mathcal{F}_{s,t}$ be the least sub-σ-field of $\mathcal{F}$ containing all the null sets and $\bigcap_{\epsilon>0} \{\Phi_{u,v} : s - \epsilon \leq u, v \leq t + \epsilon\}$. Then we have the following.

1. For $n \in \mathbb{N}$ and $x^{(1)}, \ldots, x^{(n)} \in \mathbb{R}^d$ the n-point-motion $(\phi_{s,t}(x^{(1)}), \ldots, \phi_{s,t}(x^{(n)}))$ has a Markov property with transition probabilities

$$P^{(n)}_{s,t}((x^{(1)}, \ldots, x^{(n)}), E) = \mathbb{P}\left[(\phi_{s,t}(x^{(1)}), \ldots, \phi_{s,t}(x^{(n)})) \in E\right].$$

Therein the E's are the Borel set of $\mathbb{R}^{nd}$.

2. For a $\{\mathcal{F}_{s,t} : t \in [s, \infty)\}$-stopping-time τ we have

$$\mathcal{L}\left[\Phi_{\tau,r}\left(\Phi_{s,\tau}(.)\right) : r \geq \tau \middle| \mathcal{F}_{s,\tau}\right] = \mathcal{L}_{s,\tau}\left[\Phi_{s,s+r-\tau}(.) : r \geq \tau\right].$$

Proof: [27, Theorem 4.2.1]. Note that the forward flow does not have $\mathbb{R}_+ \times \mathbb{R}_+$ as temporal index set but only $\{(s,t) \in \mathbb{R}^2 : 0 \leq s \leq t\}$. We occasionally ignore this for ease of notation because we can always add the backward flow to get the full temporal domain if necessary. □

1.2 Stochastic Flows As Random Dynamical Systems

In this section we briefly describe how to get a random dynamical system (RDS) if one already has a stochastic flow. This is necessary to be able to use concepts like stable manifolds or characteristic exponents in a precise manner. Proofs adapted to our situation can be found in some detail in [12] or more general in [1] and the references therein. We review the important definitions and the central constructions involved (and closely follow [12]). Note that to be able to view a stochastic flow as a RDS it is necessary for the flow to have stationary increments (which is perfectly satisfied in the cases we are interested in). We omit most of the proofs because they can be found in detail in [12] and [1].

1.2.1 Definitions Of The Basic Notions About RDS'

Let again $(\Omega, \mathcal{F}, \mathbb{P})$ be a probability space.

Definition 1.2.1 (two-parameter-filtration).
A two-parameter-filtration is a two-parameter-family $(\mathcal{F}_s^t : s, t \in \mathbb{R}, s \leq t)$ of sub-σ-algebras of the $\mathbb{P}$-completion $\bar{\mathcal{F}}$ of $\mathcal{F}$ such that the following holds.

1. *$\mathcal{F}_u^v \subset \mathcal{F}_s^t$ for $s \leq u \leq v \leq t$.*

2. *$\mathcal{F}_s^t = \mathcal{F}_s^{t+} := \bigcap_{u>t} \mathcal{F}_s^u$ and $\mathcal{F}_s^t = \mathcal{F}_{s-}^t := \bigcap_{u<s} \mathcal{F}_u^t$.*

3. *For any $s \leq t$ we have that $\mathcal{F}_s^t$ contains all $\mathbb{P}$-null set of $\mathcal{F}$.*

We also define $\mathcal{F}_{-\infty}^t := \bigvee_{s \leq t} \mathcal{F}_s^t$ and $\mathcal{F}_s^\infty := \bigvee_{t \geq s} \mathcal{F}_s^t$.

As usual we will omit the index sets for s and t if they are clear from the context. If one wants to form a precise picture of the object $\mathcal{F}_s^t$ one may think of $\mathcal{F}_s^t := \sigma(\phi_{u,v} : s \leq u \leq v \leq t)$ after augmentation (mod $\mathbb{P}$) and replacement by the appropriate right- (w.r.t. t) and left- (w.r.t. s) continuous versions. (see [21] for details). This means that $\mathcal{F}_s^t$ contains all the information about things taking place between time s and time t.

Definition 1.2.2 (measurable DS, MDS).
A triplet $(\Omega, \mathcal{F}, (\theta_t)_{t \in \mathbb{R}})$ with a measurable space $(\Omega, \mathcal{F})$ and a family of mappings $(\theta_t : \Omega \to \Omega : t \in \mathbb{R})$ is called a measurable dynamical system iff the following holds.

1. *The mapping $(\omega, t) \mapsto \theta_t(\omega)$ is $(\mathcal{F} \otimes \mathcal{B}(\mathbb{R}), \mathcal{F})$ measurable.*

2. *θ_0 is the identity mapping of Ω.*

3. *θ has the flow-property i.e. for any $s, t \in \mathbb{R}$ we have $\theta_{s+t} = \theta_s \circ \theta_t$.*

$(\Omega, \mathcal{F}, \mathbb{P}, (\theta_t)_{t \in \mathbb{R}})$ is called a metric dynamical system (MDS) iff the following holds.

1. *$(\Omega, \mathcal{F}, (\theta_t)_{t \in \mathbb{R}})$ is a measurable dynamical system.*

2. *$\mathbb{P}$ is a θ-invariant probability on $(\Omega, \mathcal{F})$ i.e. for any $t \in \mathbb{R}$ we have $\mathbb{P} \circ \theta_t^{-1} = \mathbb{P}$.*

A MDS with a two-parameter-filtration $\mathcal{F}_s^t$ in the sense of Definition 1.2.1 is called filtered if for any $s, t, u \in \mathbb{R}$ with $s \leq t$ we have $\theta_u^{-1} \mathcal{F}_s^t = F_{u+s}^{u+t}$.

Definition 1.2.3 (RDS).
Let $(\Omega, \mathcal{F}, \mathbb{P}, (\theta_t)_{t\in\mathbb{R}})$ be a MDS. A mapping

$$\varphi : \mathbb{R} \times \mathbb{R}^d \times \Omega \to \mathbb{R}^d$$

is called measurable random dynamical system (RDS) on (the measurable space) $(\mathbb{R}^d, \mathcal{B}(\mathbb{R}^d))$ over the MDS $(\Omega, \mathcal{F}, \mathbb{P}, (\theta_t)_{t\in\mathbb{R}})$ with time $\mathbb{R}$ iff the following conditions hold.

1. φ is $(\mathcal{B}(\mathbb{R}) \otimes \mathcal{B}(\mathbb{R}^d) \otimes \mathcal{F}, \mathcal{B}(\mathbb{R}^d))$-measurable.

2. φ satisfies the (perfect) cocycle property i.e. $\varphi(0, \omega)$ is the identity mapping of $\mathbb{R}^d$ for any $\omega \in \Omega$ and for all $s, t \in \mathbb{R}$ and $\omega \in \Omega$ we have $\varphi(t+s, \cdot, \omega) = \varphi(t, \cdot, \theta_s \omega) \circ \varphi(s, \cdot, \omega)$.

The RDS is called continuous if the mapping $(t, x) \mapsto \varphi(t, x, \omega)$ is continuous for any $\omega \in \Omega$. It is called a smooth RDS of class C^k if the mapping $x \mapsto \varphi(t, x, \omega)$ is k times continuously differentiable for any $t \in \mathbb{R}$ and $\omega \in \Omega$ and the derivatives are continuous w.r.t. (t, x). It is called linear if the mapping $x \mapsto \varphi(t, x, \omega)$ is linear for all $t \in \mathbb{R}$ and $\omega \in \Omega$.

Definition and Lemma 1.2.4 (skew-product, invariant measure).
Let φ be a RDS on $(\mathbb{R}^d, \mathcal{B}(\mathbb{R}^d))$ over the MDS $(\Omega, \mathcal{F}, \mathbb{P}, (\theta_t)_{t\in\mathbb{R}})$ with time $\mathbb{R}$. The family of mappings $(\tau_t : t \in \mathbb{R})$ defined by

$$\tau_t : \Omega \times \mathbb{R}^d \to \Omega \times \mathbb{R}^d, (\omega, x) \mapsto (\theta_t \omega, \varphi(t, x, \omega))$$

defines a measurable DS on $(\Omega \times \mathbb{R}^d, \mathcal{F} \times \mathcal{B}(\mathbb{R}^d))$. It is called the skew-product of the metric DS θ and the cocycle φ (omitting the other parts if they are clear from the context). An invariant measure for the RDS φ is a probability measure μ on $(\Omega \times \mathbb{R}^d, \mathcal{F} \otimes \mathcal{B}(\mathbb{R}^d))$ with

1. For any $t \in \mathbb{R}$ we have $\mu \circ \tau_t^{-1} = \mu$,

2. $\mu \circ \pi_\Omega^{-1} = \mathbb{P}$ where $\pi_\Omega : \Omega \times \mathbb{R}^d \to \Omega, (\omega, x) \mapsto \omega$ is the projection on Ω.

If μ is an invariant measure for φ then $(\tau_t : t \in \mathbb{R})$ is a MDS on $(\Omega \times \mathbb{R}^d, \mathcal{F} \otimes \mathcal{B}, \mu)$.

Proof: [12, p. 37]. □

We are now ready to state the multiplicative ergodic theorem in the linear case.

Theorem 1.2.5 (MET for linear RDS)**.**
Let $\varphi(t, x, \omega)$ be a linear RDS on $\mathbb{R}^d$ over the MDS $(\Omega, \mathcal{F}, \mathbb{P}, (\theta_t)_{t \in \mathbb{R}})$ such that we have that

$$\sup_{t \in [0,1]} \log^+ \|\varphi(t, \cdot)\| \text{ and } \sup_{t \in [0,1]} \log^+ \|\varphi(t, \cdot)^{-1}\|$$

belong to $L^1(\mathbb{P})$ (wherein $\|\cdot\|$ denotes the operator norm w.r.t. x). Then there exists a θ-invariant set $\tilde{\Omega} \subset \Omega$ with full $\mathbb{P}$-measure such that for all $\omega \in \tilde{\Omega}$ there is a splitting

$$\mathbb{R}^d = E_1(\omega) \oplus \ldots \oplus E_{p(\omega)}(\omega)$$

of $\mathbb{R}^d$ into random subspaces $E_i(\omega)$ of dimension $d_i(\omega)$ for $i \in \{1, \ldots, p(\omega)\}$ depending measurably on ω such that

1. *$\varphi(t, \cdot, \omega) E_i(\omega) = E_i(\theta_t \omega)$ and*

2. *$\lim_{t \to \pm \infty} \frac{1}{t} \log |\varphi(t, x, \omega)| = \lambda_i(\omega) \Leftrightarrow x \in E_i(\omega) \setminus \{0\}$.*

Moreover the numbers $\lambda_i(\omega)$, $p(\omega)$ and $d_i(\omega)$ for $i \in \{1, \ldots, p(\omega)\}$ are all θ_t-invariant for any $t \in \mathbb{R}$.

Proof: [12, Theorem 2.1.1] or [1, Chapters 3, 4 and 5]. □

The numbers $\lambda_i(\omega)$ and $d_i(\omega)$ are called the Lyapunov characteristic numbers of ϕ and their multiplicities respectively.

Definition 1.2.6 (measurable bundle)**.**
A measurable bundle (Y, Ω, π) with typical fibre X consists of a measurable space $(Y, \mathcal{Y})$, a measurable space $(\Omega, \mathcal{F})$, whose one-point-sets are measurable, a measurable space $(X, \mathcal{B})$, a measurable onto map $\pi: Y \to \Omega$ and a global measurable trivialization i.e. a bimeasurable bijection $\Psi: Y \to \Omega \times X$ such that $\pi_\Omega \circ \Psi = \pi$ (π_Ω is again the projection $(\omega, x) \mapsto \omega$). In particular for any $\omega \in \Omega$ the mapping

$$\psi(\omega) := \Psi \mid_{\pi^{-1}(\{\omega\})}: \pi^{-1}(\{\omega\}) \to \{\omega\} \times X$$

is a bimeasurable bijection w.r.t. the corresponding trace-σ-algebras.
A measurable bundle is called linear if the typical fibre X and all the fibres $\pi^{-1}(\{\omega\})$ have the structure of a d-dimensional (real) vector space and ψ is linear in the sense that $\pi_X \circ \psi(\omega): \pi^{-1}(\{\omega\}) \to X$ is linear for all $\omega \in \Omega$.

Definition 1.2.7 (bundle RDS).
Let $(\Omega, \mathcal{F}, \mathbb{P}, (\theta_t)_{t \in \mathbb{R}})$ be a MDS and let (Y, Ω, π) be a measurable bundle with typical fibre X. A measurable bundle RDS over $(\Omega, \mathcal{F}, \mathbb{P}, (\theta_t)_{t \in \mathbb{R}})$ is a measurable DS τ on $(Y, \mathcal{Y})$ which preserves fibres i.e.

$$\pi \circ \tau_t = \theta_t \circ \pi$$

The latter is equivalent to the statement that the fibre mappings

$$\varphi(t, \cdot, \omega) : \tau_t \mid_{\pi^{-1}(\{\omega\})} \colon \pi^{-1}(\{\omega\}) \to \pi^{-1}(\{\theta_t \omega\})$$

form a cocycle over θ i.e. $\varphi(0, \cdot, \omega)$ is the identity of $\pi^{-1}(\{\omega\})$ and for $t, s \in \mathbb{R}$ we have $\varphi(t+s, \cdot, \omega) = \varphi(t, \cdot, \theta_s \omega) \circ \varphi(s, \cdot, \omega)$. If the fibre mappings $\varphi(t, \cdot, \omega)$ are linear for any $t \in \mathbb{R}$ and $\omega \in \Omega$ then the bundle RDS is called linear.

Proposition 1.2.8 (differential as bundle RDS).
Let φ be a RDS of class C^1 on a d-dimensional manifold M over the MDS $(\Omega, \mathcal{F}, \mathbb{P}, (\theta_t)_{t \in \mathbb{R}})$ with time $\mathbb{R}$ and invariant measure μ. Then

1. *The differential $D\varphi(t, \cdot, \omega) : TM \to TM, (x, v) \mapsto (\varphi(t, x, \omega), D\varphi(t, x, \omega)v)$ is a continuous cocycle where of course TM denotes the tangent bundle of M.*

2. *The family $(T(t) : t \in \mathbb{R})$ defined by $T(t) : \Omega \times TM \to \Omega \times TM, (\omega, (x, v)) \mapsto (\theta_t \omega, D_x \varphi(t, \cdot, \omega)v)$ is a linear bundle RDS on $\Omega \times TM$ over the skew-product MDS*

$$\tau_t : \Omega \times TM \to \Omega \times TM \text{ with } \tau_t(\omega, x) = (\theta_t \omega, \varphi(t, x, \omega)).$$

Proof: [12, Proposition 2.1.1]. Note that the proof does not depend on the fact that μ is a probability so we can use the proposition for σ-finite measures as invariant measures μ. □

Theorem 1.2.9 (MET for linearized smooth cocycles).
Let $\varphi(t, x, \omega)$ be a C^1-RDS on a d-dimensional manifold M over the MDS $(\Omega, \mathcal{F}, \mathbb{P}, (\theta_t)_{t \in \mathbb{R}})$ with invariant probability μ. Consider the linear bundle RDS T (see Proposition 1.2.8) on $\Omega \times TM$ over the MDS $(\Omega \times M, \mathcal{F} \otimes \mathcal{B}(M), \mu, (\tau_t)_{t \in \mathbb{R}})$ with τ_t as defined in Proposition 1.2.8.

Assume further the following integrability conditions i.e. assume that

$$\sup_{t\in[0,1]} \log^+ \|D_x\varphi(t,\cdot,\omega)(\cdot)\| \quad \text{and} \quad \sup_{t\in[0,1]} \log^+ \|D_x\varphi(t,\cdot,\omega)^{-1}(\cdot)\|$$

belong to $L^1(\mu)$ where $\|\cdot\|$ denotes the corresponding operator norms of the differential as a linear mapping between the tangent spaces at the appropriate points in M. Then there exists a τ-invariant set $\Delta \subset \Omega \times M$ with full μ-measure such that for all $(\omega,x) \in \Delta$ there is a splitting

$$T_xM = E_1(\omega,x) \oplus \ldots \oplus E_{p(\omega,x)}(\omega)$$

of T_xM into random subspaces $E_i(\omega,x)$ of dimension $d_i(\omega)$ for $i \in \{1,\ldots,p(\omega)\}$ depending measurably on ω such that for $i \in \{1,\ldots,p(\omega,x)\}$ we have

1. $D_x\varphi(t,\cdot,\omega)E_i(\omega,x) = E_i(\tau_t(\omega,x)) = E_i(\theta_t\omega, \varphi(t,x,\omega))$,

2. $\lim_{t\to\pm\infty} \frac{1}{t} \log |D_x\varphi(t,\cdot,\omega)v| = \lambda_i(\omega,x) \Leftrightarrow v \in E_i(\omega,x) \setminus \{0\}$.

Moreover the numbers $\lambda_i(\omega,x)$, $p(\omega,x)$ and $d_i(\omega,x)$ for $i \in \{1,\ldots,p(\omega,x)\}$ are all θ_t-invariant for any $t \in \mathbb{R}$.

Proof: [1, Theorem 4.2.6]. A similar situation in the discrete time case is treated in [24, Theorem 3.1.1 and Theorem 5.1.1]. □

1.2.2 Construction Of A RDS From A Stochastic Flow

We are now ready to indicate how one obtains a RDS from a stochastic flow. This has been done in [2]. To be precise, we sketch the solution to the following problem.

Problem 1.2.10 (stochastic flows as RDS).
Given a stochastic flow $(\phi_{s,t}(\cdot) : s,t \geq 0)$ with stationary increments, we have to find a MDS $(\Omega, \mathcal{F}, \mathbb{P}, (\theta_t)_{t\in\mathbb{R}})$ and a RDS φ over $(\Omega, \mathcal{F}, \mathbb{P}, (\theta_t)_{t\in\mathbb{R}})$ such that

$$\mathcal{L}\left[(\varphi(t-s,x,\theta_s\cdot) : s,t \geq 0, x \in \mathbb{R}^d)\right] = \mathcal{L}\left[(\phi_{s,t}(x,\cdot) : s,t \geq 0, x \in \mathbb{R}^d)\right].$$

The solution is originally due to L. Arnold and M. Scheutzow (see [2]) with slightly different assumptions. For their proof is perfectly valid in our setting (which has been noticed e.g. in [12]) we do not go into all the details but we nevertheless state the precise result.

Theorem 1.2.11 (stochastic flows as RDS).
Let $(\phi_{s,t} : s, t \geq 0)$ be a stochastic flow of diffeomorphisms defined on a probability space $(\Omega, \mathcal{F}, \mathbb{P})$ such that its forward component $(\phi_{s,t} : 0 \leq s \leq t)$ is generated by the semimartingale field $F(t, x, \omega) : \mathbb{R}_+ \times \mathbb{R}^d \times \Omega \to \mathbb{R}^d$ via the Kunita-type Stratonovich SDE

$$\phi_{s,t}(x) = x + \int_s^t F(\circ du, \phi_{s,u}(x)) \text{ for all } 0 \leq s \leq t < \infty \text{ and } x \in \mathbb{R}^d.$$

Assume that F is a $C^{k,\delta}$-semimartingale field with local characteristics (a, b) belonging to the class $(B_b^{k+1,\delta}, B_b^{k,\delta})$ for some $k \geq 1$ and $\delta \geq 0$ such that

$$\sum_{j=1}^d \frac{\partial a^{\cdot j}}{\partial x_j}(x, y, t)\mid_{x=y}$$

belongs to the class $B_b^{k,\delta}$. Assume further that F has stationary increments. Then there exists a filtered MDS $(\tilde{\Omega}, \tilde{\mathcal{F}}, (\tilde{\mathcal{F}}_s^t) : s \leq t \in \mathbb{R}), \tilde{\mathbb{P}}, (\theta_t)_{t \in \mathbb{R}})$ and for $0 \leq \epsilon < \delta$ a RDS of class $C^{k,\epsilon}$ called $\varphi : \mathbb{R} \times \mathbb{R}^d \times \tilde{\Omega} \to \mathbb{R}^d$ over $(\tilde{\Omega}, \tilde{\mathcal{F}}, \tilde{\mathbb{P}}, (\theta_t)_{t \in \mathbb{R}})$ such that the distribution of $(\phi_{s,t}(x, \cdot) : s, t \geq 0, x \in \mathbb{R}^d)$ coincides with the distribution of $(\varphi(t - s, x, \theta_s \cdot) : s, t \geq 0, x \in \mathbb{R}^d)$.

Sketch of proof: (a complete proof adapted to our situation can be found in [2] or adapted to the case of an IOUF in [12]) The construction is as follows.
Let $\tilde{\Omega} := \{f : \mathbb{R} \to C(\mathbb{R}^d, \mathbb{R}^d) : f \text{ is continuous and } f(0) = 0\}$ wherein we take on $C(\mathbb{R}^d, \mathbb{R}^d)$ and on $\tilde{\Omega}$ the topology of uniform convergence on compacts. Let $\tilde{\mathcal{F}} := \mathcal{B}(\tilde{\Omega})$ and define the probability measure $\tilde{\mathbb{P}}$ by fixing the finite-dimensional distributions as follows. For $n \in \mathbb{N}$ and $t_1 < \ldots < t_n$ and $A \in \mathcal{B}(C(\mathbb{R}^d, \mathbb{R}^d))^{\otimes n}$ let

$$\tilde{\mathbb{P}}\left[(\tilde{\omega}(t_n) - \tilde{\omega}(t_{n-1}), \ldots, \tilde{\omega}(t_2) - \tilde{\omega}(t_1)) \in A\right]$$
$$= \mathbb{P}\left[F(t_n - t_1, \cdot) - F(t_{n-1} - t_1, \cdot), \ldots, F(t_2 - t_1, \cdot) - F(0, \cdot) \in A\right].$$

After restricting $\tilde{\mathbb{P}}$ to $\tilde{\Omega}$ and defining the mappings $\theta_t : \tilde{\Omega} \to \tilde{\Omega}$ for $t \in \mathbb{R}$ by

$$\theta_t \tilde{\omega}(s) = \tilde{\omega}(t + s) - \tilde{\omega}(t) \tag{1.3}$$

1.2 Stochastic Flows As Random Dynamical Systems

as well as the two-parameter-filtration

$$\tilde{\mathcal{F}}_s^t := \bigcap_{\epsilon > 0} \sigma\{\tilde{\omega}(u) - \tilde{\omega}(v) : s - \epsilon \leq u \leq v \leq t + \epsilon\}$$

(and $\tilde{\mathbb{P}}$-augmentation) we end up with a filtered MDS in the sense of Definition 1.2.2. Introducing the semimartingale field $\tilde{F}(t, x, \tilde{\omega}) := \tilde{\omega}(t, x)$ one gets a stochastic flows of diffeomorphisms $\tilde{\phi}_{s,t}$ as the solution to the SDE

$$\tilde{\phi}_{s,t}(x) - x = \begin{cases} \int_s^t \tilde{F}(\circ du, \tilde{\phi}_{s,u}(x)) & : s \leq t \\ -\int_t^s \tilde{F}(\circ du, \tilde{\phi}_{s,u}(x)) & : t \leq s \end{cases}.$$

Finally letting

$$\varphi : \mathbb{R} \times \mathbb{R}^d \times \tilde{\Omega} \to \mathbb{R}^d, (t, x, \tilde{\omega}) \mapsto \tilde{\phi}_{0,t}(x, \tilde{\omega})$$

one shows that this φ can be modified to satisfy all the needs (we again do not indicate the modification in the notation). □

Note that the difference between the Kunita-type Itô-SDE and the Kunita-type Stratonovich-SDE does not play any role at all (since we might rewrite an SDE given in one type into the other one and vice versa).

Proposition 1.2.12 (ergodicity of the RDS).
In the setting of Theorem 1.2.11 assume further that F has independent increments. Then the filtered MDS $(\tilde{\Omega}, \tilde{\mathcal{F}}, (\tilde{\mathcal{F}}_s^t : s \leq t \in \mathbb{R}), \tilde{\mathbb{P}}, (\theta_t)_{t \in \mathbb{R}})$ is ergodic.

Proof: [12, Proposition 2.2.2]. □

1.2.3 Time Discretization, Markov Chain Representations

The previous constructions have all been carried out for continuous time, but we will need discrete time versions of them. Restricting $(\tilde{\Omega}, \tilde{\mathcal{F}}, (\tilde{\mathcal{F}}_s^t : s \leq t \in \mathbb{R}), \tilde{\mathbb{P}}, (\theta_t)_{t \in \mathbb{R}})$ and $\varphi : \mathbb{R} \times \mathbb{R}^d \times \tilde{\Omega} \to \mathbb{R}^d$ from Theorem 1.2.11 to $(\tilde{\Omega}, \tilde{\mathcal{F}}, (\tilde{\mathcal{F}}_s^t : s \leq t \in \mathbb{N}), \tilde{\mathbb{P}}, (\theta_t)_{t \in \mathbb{N}})$ and $\varphi : \mathbb{N} \times \mathbb{R}^d \times \tilde{\Omega} \to \mathbb{R}^d$ we obtain the appropriate discretization we will be working with occasionally (especially in Chapter 6). Note that all the previous results are still valid in discrete time (with the obvious modifications). The analysis we will be developing applies to the composition of a sequence of i.i.d. random transformations. To be precise we have the following generic Markov chain version of our setting. Let μ be an arbitrary measure on $\mathbb{R}^d$ and let $x \in \mathbb{R}^d$ be distributed according to μ independent of φ. Then $X_t := \varphi(t,x)$ for $t \in \mathbb{N}$ defines a Markov chain in discrete time with state space $\mathbb{R}^d$.

Remark 1.2.13 (Kifer's setting).
Having the discrete time RDS $\varphi : \mathbb{N} \times \mathbb{R}^d \times \tilde{\Omega} \to \mathbb{R}^d$ over $(\tilde{\Omega}, \tilde{\mathcal{F}}, \tilde{\mathbb{P}}, (\theta_t)_{t\in\mathbb{N}})$ constructed from a Brownian flow with stationary increments we get a sequence $\phi_{0,1}, \phi_{1,2}, \ldots$ of i.i.d. $\mathrm{Diff}(\mathbb{R}^d)$-valued random variables defining $\phi_{i,i+1}(\tilde{\omega}) := \varphi(1,\cdot,\theta^i\tilde{\omega})$ and the Markov chain mentioned above can be written as $X_t = (\phi_{t,t-1}(\tilde{\omega}) \circ \ldots \circ \phi_{0,1}(\tilde{\omega}))(x)$ which (in law) is nothing but the unitstep-discretization of the one-point-motion of the flow we were starting with.

We are now ready to state the result on invariant measures we already announced.

Lemma 1.2.14 (different notions of invariant measures).
Let ϕ be a stochastic flow with stationary increments and φ be the associated RDS with the notation of Theorem 1.2.11. The probability measure μ on $\mathbb{R}^d$ is invariant for the one-point-motion of ϕ (in the sense of discrete time Markov chains) iff the measure $\mu \otimes \tilde{\mathbb{P}}$ restricted to discrete time is invariant for φ.

Proof: [42] or [24, Lemma 1.2.3]. □

We will identify a stochastic flow with its associated RDS where possible, e.g. we will speak about the entropy of ϕ or about its characteristic exponents.

1.3 Isotropic Brownian Flows

The transition from a general stochastic flow of homeomorphisms to an isotropic Brownian flow (IBF) is performed by specifying its Itô's random infinitesimal generator i.e. the semimartingale field driving the flow which in this case turns out to be a martingale field. We briefly describe the construction of this field in the sequel. See [28] or [5] for further details. One can obtain an IBF by letting $b(x,t) \equiv 0$ and $a(x,y,t) = tb(x-y)$ for a suitable function $b : \mathbb{R}^d \to \mathbb{R}^{d \times d}$. This function ist the so-called isotropic covariance tensor, which we are about to define precisely. Do not mix the covariance tensor $b(x)$ of an IBF with the drift-part $b(x,t)$ of the local characteristic of a general stochastic flow

1.3 ISOTROPIC BROWNIAN FLOWS

(which in the IBF-case vanishes). We do not want to rename one of the b's because both notations are very established in the literature. IBFs have been considered among others by [17], [5], [28], [9], [36], [45], [54], [12] and [56].

1.3.1 Covariance Tensors And Correlation Functions

Definition 1.3.1 (isotropic covariance tensor).
A function $b : \mathbb{R}^d \to \mathbb{R}^{d \times d}$ is an isotropic covariance tensor if the following holds true.

1. *$x \mapsto b(x)$ is C^4 and all derivatives up to order four are bounded.*

2. *$b(0) = E_d$ (the d-dimensional identity matrix).*

3. *$x \mapsto b(x)$ is not constant.*

4. *$b(x) = O^* b(Ox) O$ for any $x \in \mathbb{R}^d$ and any orthogonal matrix $O \in O(d)$.*

The assumptions on the differentiability of the generating tensor b are a bit restrictive, but we do not want to mess with smoothness problems. Occasionally we may require b to be much more than C^4 if this seems to be a reasonable way to overcome technical difficulties, that do not have anything to do with the „real" problem we work on.

Definition and Lemma 1.3.2 (correlation functions and covariance tensors).
Let b be an isotropic covariance tensor. The functions

$$B_L(r) := b_{ii}(re_i), r \geq 0$$

and

$$B_N(r) := b_{ii}(re_j), r \geq 0, i \neq j$$

are the longitudinal (and normal - respectively) correlation functions of b. Their definitions do not depend on the specific choice of $1 \leq i, j \leq d$ and we have for arbitrary $i, j \in \{1, \ldots, d\}$ and $x \in \mathbb{R}^d$ that

$$b^{ij}(x) = \begin{cases} (B_L(|x|) - B_N(|x|)) x^i x^j / |x|^2 + B_N(|x|) \delta^{ij} & : x \neq 0 \\ \delta^{ij} = \delta^{ij} B_L(0) = \delta^{ij} B_N(0) & : x = 0 \end{cases}.$$

The following holds (notation as above).

$$\beta_L := -B_L''(0) > 0,$$
$$\beta_N := -B_N''(0) > 0,$$
$$B_L(r) = 1 - \frac{1}{2}\beta_L r^2 + O(r^4) : (r \to 0), \tag{1.4}$$
$$B_N(r) = 1 - \frac{1}{2}\beta_N r^2 + O(r^4) : (r \to 0), \tag{1.5}$$
$$\|B_L\|_\infty \vee \|B_N\|_\infty \leq 1,$$
$$\lim_{|x|\to\infty} b(x) = 0, \tag{1.6}$$

$$\forall \epsilon > 0 : \exists r^{(\epsilon)} > 0 : \forall r > r^{(\epsilon)} : |B_L(r)| \vee |B_N(r)| < 1 - \epsilon.$$

The partial derivatives of b at 0 satisfy $\partial_k \partial_l b^{i,j}(0) = \frac{1}{2}(\beta_N - \beta_L)(\delta_{ki}\delta_{lj} + \delta_{kj}\delta_{li}) - \beta_N \delta_{kl}\delta_{ij}$ and of course $\partial_k \partial_l b^{i,i}(0) = (\beta_N - \beta_L)\delta_{ki}\delta_{li} - \beta_N \delta_{kl}$.

Proof: [5, (2.5), (2.6), (2.8), (2.9), (2.18) and the discussion after (2.13)]. □

Note that not all functions can be used as candidates for B_N and B_L (even if they have the properties given above). In [60] one can find a parameterization of all isotropic covariance tensors in terms of two finite measures on $(0, \infty)$ as integral transforms involving some Bessel functions. If one wants to have a complete overview over the possible tensors one should start with these measures and obtain the correlation functions from them. Since we do not need this we do not give more details. However observe that we implicitly made several normalizing assumptions for the ease of notation which we explain in the sequel: First one can add a constant times the identity to b without changing any of the properties appearing in the definition of „isotropic covariance tensor". This corresponds to the addition of a rigid translation of the space by a Brownian motion to the generated IBF. For the properties we are interested in do not depend on this translation we assume it to be zero (so (1.6) is the normalizing assumption to ensure this). The fact that $B_L(0) = B_N(0) = 1$ has to be seen in a similar way. Suppose we have a stochastic flow written as $(\phi_t : t \geq 0)$ which has some kind of invariance property (think of the invariance properties of an IBF we give later). Then usually for $K > 0$ we have that $(\phi_{Kt} : t \geq 0)$ is also a stochastic flow with the same invariance property. The choice $B_L(0) = B_N(0) = 1$ ensures that the one-point-motion of an IBF is a standard Brownian motion (other choices lead to stochastic processes that are constant multiples of Brownian motions). Basically one can put it this way: if one wants to have a translation-free IBF on the standard time scale, then the isotropic covariance tensors to consider are exactly the ones given above. We will always assume this when speaking of an isotropic covariance tensor.

Remark: We also give the statements (1.4) and (1.5) in the following weaker forms:

$$\forall \epsilon > 0 : \exists r^{(\epsilon)} > 0 : \forall r < r^{(\epsilon)} : |1 - B_L(r) - \frac{1}{2}\beta_L r^2| \vee |1 - B_N(r) - \frac{1}{2}\beta_N r^2| < \epsilon r^3. \quad (1.7)$$

$$\exists 0 \leq \bar{r} \leq 1, C > 0 : \forall |x| \leq \bar{r} : |\partial_k \partial_l b^{i,j}(x) - \partial_k \partial_l b^{i,j}(0)| \leq C|x|^2. \quad (1.8)$$

1.3.2 Brownian Fields And Generated Flows

Now we can define the semimartingale field F which in fact coincides with its martingale part M. More precisely it coincides with its local martingale part and this local martingale part is in fact a true martingale field.

Definition 1.3.3 (isotropic Brownian field, IBF).
Let b be an isotropic covariance tensor. A random vector field $\left(M(t,x) : t \geq 0, x \in \mathbb{R}^d\right)$ with values in $\mathbb{R}^d$ - defined on a probability space $(\Omega, \mathcal{F}, \mathbb{P})$ - is an isotropic Brownian field if the following holds.

1. *$(t, x) \mapsto M(t, x)$ is a centered Gaussian process.*

2. *$\text{cov}(M(s, x), M(t, y)) = (s \wedge t) b(x - y)$.*

3. *$(t, x) \mapsto M(t, x)$ is continuous for almost all ω.*

The existence of such a field follows from Kolmogorov's Existence Theorem and Theorem 1.1.1. A stochastic flow, generated via (1.2) with $F(t,x) = M(t,x)$ (i.e. $V(t,x) \equiv 0$) for an isotropic Brownian field $M(t,x)$ is called an isotropic Brownian flow (IBF).

We state some useful properties of isotropic Brownian fields.

Lemma 1.3.4 (properties of isotropic Brownian fields).
Isotropic Brownian fields fulfil the following.

1. *$t \mapsto M(t, x)$ is a d-dimensional standard Brownian motion.*

2. $\langle M(.,x), M(.,y)\rangle_t = b(x-y)t$ for $x, y \in \mathbb{R}^d$ and $t \in [0, \infty)$ as well as

$$\left\langle \partial_l M^i(\cdot, x.), \partial_k M^j(\cdot, y.)\right\rangle_t = -\int_0^t \partial_l \partial_k b^{i,j}(x_s - y_s) ds. \tag{1.9}$$

Let b be an isotropic covariance tensor. For an $\mathbb{R}^d$-valued Gaussian field $U = \left(U(x) : x \in \mathbb{R}^d\right)$ with covariance $\mathrm{cov}(U(x+y), U(y)) = b(x)$ the following holds.

3. $U(x)$ has a differentiable modification.

4. $\beta_L = \mathbb{E}\left[\frac{\partial U_i}{\partial x_i}\right]^2$ and $\beta_N = \mathbb{E}\left[\frac{\partial U_i}{\partial x_j}\right]^2$ for $i \neq j$.

Proof: 1., 2. are consequences of Definition 1.3.3. 3.: [27, Theorem 1.4.1]. 4.: [5, (2.7)]. □

Remark: The above especially applies to $U(x) = M(1, x)$. In the following we list some properties of IBFs which follow more or less directly from the definitions and which give the reason why the IBFs are the flow versions of Brownian motions in some sense. First we state some global properties of their law (as G-valued random variables).

Lemma 1.3.5 (general properties of IBFs).
Let $(\Phi_{s,t} : 0 \leq s, t < \infty)$ be an IBF. Then it is a Brownian flow that satisfies the following.

1. It is temporally homogenous i.e. for $C > 0$ the laws of $(\Phi_{s,t} : 0 \leq s, t < \infty)$ and $(\Phi_{s+C,t+C} : 0 \leq s, t < \infty)$ coincide.

2. It is rotation invariant i.e. for any orthogonal matrix O we have that $(O\Phi_{s,t}(\cdot) : 0 \leq s, t < \infty)$ and $(\Phi_{s,t}(O\cdot) : 0 \leq s, t < \infty)$ coincide in law.

3. It is spatially homogenous (or translation invariant) i.e. for any $x \in \mathbb{R}^d$ we have that $(\Phi_{s,t} : 0 \leq s, t < \infty)$ and $(\Phi_{s,t}(\cdot + x) : 0 \leq s, t < \infty)$ coincide in law.

4. It has independent increments i.e. for $0 < t_1 < \ldots < t_n < \infty$ we have that the family of random mappings $(\Phi_{0,t_1}, \Phi_{t_1,t_2}, \ldots, \Phi_{t_{n-1},t_n})$ is independent.

Proof: [12, Section 1.2]. □

The latter means roughly that sitting on a particle which is subject to an IBF one can neither determine where one is nor the direction one is looking at by observation of the flow. This property essentially implies quite a lot of structure for the finite-dimensional motions of an IBF some of which we state in the following theorem. See [5] or [28] for more details.

1.3 ISOTROPIC BROWNIAN FLOWS

Theorem 1.3.6 (finite-dimensional diffusions).
Let $(\Phi_{s,t} : 0 \leq s, t < \infty)$ be an IBF. We have the following properties for the n-point-motion $(x_t^{(1)}, \ldots, x_t^{(n)}) := (\Phi_t(x^{(1)}), \ldots, \Phi_t(x^{(n)}))$.

1. It is an $\mathbb{R}^{nd}$-valued diffusion, with generator $L^{(n)}$ given for $g \in C_b^2$ by

$$L^{(n)}g\left(x^{(1)}, \ldots, x^{(n)}\right) = \frac{1}{2} \sum_{l,m=1}^{n} \sum_{i,j=1}^{d} b\left(x^{(l)} - x^{(m)}\right) \frac{\partial^2 g}{\partial x_i^{(l)} \partial x_j^{(m)}} \left(x^{(1)}, \ldots, x^{(n)}\right). \quad (1.10)$$

2. $\min_{l \neq m} \left\| x^{(l)} - x^{(m)} \right\| \to \infty \Rightarrow |L^{(n)}g\left(x^{(1)}, \ldots, x^{(n)}\right) - \frac{1}{2}\Delta g\left(x^{(1)}, \ldots, x^{(n)}\right)| \to 0$, where $\frac{1}{2}\Delta$ is the generator of a nd-dimensional Brownian motion.

3. $\rho_t^{xy} := \|x_t - y_t\|$ is a diffusion on $(0, \infty)$ with generator $\bar{L}$ given for $g \in C_b^2$ by

$$\bar{L}g(r) = (d-1)\left(\frac{1 - B_N(r)}{r}\right) g'(r) + (1 - B_L(r))g''(r). \quad (1.11)$$

4. ρ_t^{xy} solves the SDE

$$d\rho_t^{xy} = (d-1)\left(\frac{1 - B_N(\rho_t^{xy})}{\rho_t^{xy}}\right) dt + \sqrt{2(1 - B_L(\rho_t^{xy}))} dW_t \quad (1.12)$$

with a standard Brownian motion $(W_t)_{t \geq 0}$.

The spatial derivative $D_x \Phi$ solves the SDE

$$D_x \Phi_{0,t}(\cdot) = \text{id}_{\mathbb{R}^d} + \int_0^t D_{\Phi_{0,u}(\cdot)} M(du, \Phi_{0,u}(\cdot)) D_x \Phi_{0,u}(\cdot). \quad (1.13)$$

Proof: [28, p. 617], [27, p.124], [28, p. 4] and [5, (3.11)]. □

Remark: [28] uses a slightly different definition. Assume $\alpha = 1$ there to get things into line with the definitions above (see the discussion after Definition and Lemma 1.3.2). [45] considers the mean square separation of $(\rho_t^{xy})^2$ and derives precise asymptotics for it. The previous theorem shows, that for $n = 2$ we get that (x_t, y_t) coincides in law with the solution of the following SDE.

$$\begin{pmatrix} x_t \\ y_t \end{pmatrix} - \begin{pmatrix} x \\ y \end{pmatrix} = \int_0^t \begin{pmatrix} E_d & b(x_s - y_s) \\ b(x_s - y_s) & E_d \end{pmatrix}^{1/2} dW_s =: \int_0^t \bar{b}(x_s - y_s) dW_s. \quad (1.14)$$

Therein $(W_t)_{t\geq 0}$ is a $2d$-dimensional standard Brownian motion. The following lemma states some information about the eigenvalues of b and $\bar{b}$ respectively.

Lemma 1.3.7 (eigenvalues of b and $\bar{b}$).
For $z \in \mathbb{R}^d$ we have:

1. *z is an eigenvector of $b(z)$ to the eigenvalue $B_L(|z|)$.*

2. *Any vector $0 \neq z^\perp$ perpendicular to z is an eigenvector of $b(z)$ to the eigenvalue $B_N(|z|)$.*

3. *$\bar{b}$ has the eigenvalues $\{1 \pm B_L(z), 1 \pm B_N(z)\}$ with multiplicities 1 and $d-1$ respectively.*

Proof: Since 1. and 2. have also been shown in [25] (for general d) we only give their proof in the two-dimensional case. The general one is perfectly similar. In the following computations we omit the arguments of B_L and B_N (so B_L means $B_L(|z|)$.)
1.: For $z = (z_1, z_2)$ we have (see Lemma 1.3.2) that

$$
\begin{aligned}
b(z)z &= \begin{pmatrix} \frac{(B_L-B_N)z_1^2}{\|z\|^2} + B_N & \frac{(B_L-B_N)z_1 z_2}{\|z\|^2} \\ \frac{(B_L-B_N)z_1 z_2}{\|z\|^2} & \frac{(B_L-B_N)z_2^2}{\|z\|^2} + B_N \end{pmatrix} \begin{pmatrix} z_1 \\ z_2 \end{pmatrix} \\
&= \frac{1}{\|z\|^2} \begin{pmatrix} B_L z_1^2 + B_N z_2^2 & (B_L - B_N)z_1 z_2 \\ (B_L - B_N)z_1 z_2 & B_L z_2^2 + B_N z_1^2 \end{pmatrix} \begin{pmatrix} z_1 \\ z_2 \end{pmatrix} \\
&= \frac{1}{\|z\|^2} \begin{pmatrix} B_L z_1^3 + B_N z_1 z_2^2 + B_L z_1 z_2^2 - B_N z_1 z_2^2 \\ B_L z_1^2 z_2 - B_N z_1^2 z_2 + B_L z_2^3 + B_N z_1^2 z_2 \end{pmatrix} \\
&= B_L z.
\end{aligned}
$$

2.: W.l.o.g. we let $z^\perp := (z_2, -z_1)^T$ and conclude

$$
\begin{aligned}
b(z)z^\perp &= \frac{1}{\|z\|^2} \begin{pmatrix} B_L z_1^2 + B_N z_2^2 & (B_L - B_N)z_1 z_2 \\ (B_L - B_N)z_1 z_2 & B_L z_2^2 + B_N z_1^2 \end{pmatrix} \begin{pmatrix} z_2 \\ -z_1 \end{pmatrix} \\
&= \frac{1}{\|z\|^2} \begin{pmatrix} B_L z_1^2 z_2 + B_N z_2^3 - B_L z_1^2 z_2 + B_N z_1^2 z_2 \\ B_L z_1 z_2^2 - B_N z_1 z_2^2 - B_L z_1 z_2^2 - B_N z_1^3 \end{pmatrix} \\
&= B_N z^\perp.
\end{aligned}
$$

3. follows from 1. and 2. with the following simple computations (valid for general d).

$$\begin{pmatrix} E_2 & b(z) \\ b(z) & E_2 \end{pmatrix} \begin{pmatrix} z \\ z \end{pmatrix} = (1 + B_L) \begin{pmatrix} z \\ z \end{pmatrix},$$

$$\begin{pmatrix} E_2 & b(z) \\ b(z) & E_2 \end{pmatrix} \begin{pmatrix} -z \\ z \end{pmatrix} = (1 - B_L) \begin{pmatrix} -z \\ z \end{pmatrix},$$

$$\begin{pmatrix} E_2 & b(z) \\ b(z) & E_2 \end{pmatrix} \begin{pmatrix} z^\perp \\ z^\perp \end{pmatrix} = (1 + B_N) \begin{pmatrix} z^\perp \\ z^\perp \end{pmatrix},$$

$$\begin{pmatrix} E_2 & b(z) \\ b(z) & E_2 \end{pmatrix} \begin{pmatrix} z^\perp \\ -z^\perp \end{pmatrix} = (1 - B_N) \begin{pmatrix} z^\perp \\ -z^\perp \end{pmatrix}.$$

□

Observe that the previous lemma ensures that $\bar{b}$ is elliptic apart from the diagonal $\{x = y\}$. The regularity of the finite-dimensional projections given above can also be used in terms of the running maximum of a geometric Brownian motion.

Lemma 1.3.8 (Two-Point-Control - IBF version).
Let $(\Phi_{s,t} : 0 \leq s \leq t < \infty)$ be an IBF. There are constants $\lambda > 0$ and $\bar{\sigma} > 0$ such that for $x, y \in \mathbb{R}^d$ there is a standard Brownian motion $(W_t)_{t \geq 0}$ such that we have for $(x_t, y_t) := (\Phi_t(x), \Phi_t(y))$ the following.

1. *We have a.s. for all $t \geq 0$ that*

$$|x_t - y_t| \leq |x - y| e^{\bar{\sigma} \sup_{0 \leq s \leq t} W_s + \lambda t}. \tag{1.15}$$

2. *We have for each $x, y \in \mathbb{R}^d$, $T > 0$ and $q \geq 1$ that*

$$\mathbb{E}\left[\sup_{0 \leq t \leq T} |x_t - y_t|^q\right]^{1/q} \leq 2|x - y| e^{(\lambda + \frac{1}{2} q \bar{\sigma}^2) T}. \tag{1.16}$$

Proof: [52, Condition (H) and Lemma 4.1]. □

Observe that we write $\mathbb{E}[X]^q$ for $(\mathbb{E}[X])^q$ which is different from $\mathbb{E}[X^q]$. We will use this convention throughout the whole work.

1.3.3 Lyapunov Exponents

Lyapunov exponents are usually obtained for linear RDS from the multiplicative ergodic theorem (Theorem 1.2.9) if the RDS has an invariant probability measure. So to define Lyapunov exponents for an IBF we might try to find a RDS that naturally corresponds to the IBF, hope that this RDS preserves some nice probability measure and apply the standard theory which we briefly sketched in Section 1.2 and e.g. can be found in detail in [1]. The difficulties arising especially from null sets in constructing the RDS have been overcome in [2] as already seen (see Theorem 1.2.11) and we do not go into details here. Since RDS' and stochastic flows with stationary increments are essentially the same objects one indeed can find such an RDS which unfortunately has no invariant probability but Lebesgue measure as invariant measure (which one should expect in the first place according to the spatial invariance properties of IBFs). To overcome also this issue one might center the flow around some trajectory i.e. pass to $\Phi_{s,t}(\cdot) - \Phi_{s,t}(x) + x$ for some $x \in \mathbb{R}^d$. In this way one gets a RDS with an invariant probability which essentially makes the multiplicative ergodic theorem applicable to IBFs. The other possibility is to observe that for any $x \in \mathbb{R}^d$ the spatial derivative $D_x\Phi_{0,n}(\cdot)$ coincides in law with the law of the product of n i.i.d. random variables each having the distribution of $D_x\Phi_{0,1}(\cdot)$. This of course can be realised as a linear RDS which delivers an approach that does not rely on any invariant probabilities. The Lyapunov spectrum of an IBF has been computed in [5]. For a more detailed account on the issues arising on the way to the following theorem see [12].

Theorem 1.3.9 (Lyapunov exponents for IBFs).
Let $(\Phi_{s,t} : 0 \leq s, t < \infty)$ be an IBF with covariance tensor b. Then for $\lambda \otimes \mathbb{P}$-almost all $(x, \omega) \in \mathbb{R}^d \times \Omega$ there is a measurable (i.e. random) family of linear vectorspaces $V_d(x,\omega) \subset \ldots \subset V_1(x,\omega)$ with $\dim(V_i) = d + 1 - i$ for $i = 1, \ldots, d$ such that

$$\lim_{n \to \infty} \frac{1}{n} \log |D_x\Phi_n(\omega, x)v| = \mu_i \Leftrightarrow v \in V_i(x,\omega) \setminus V_{i+1}(x,\omega)$$

wherein $(\mu_1 > \ldots > \mu_d)$ are constants (neither depending on x nor on ω). The numbers μ_i are called the Lyapunov exponents of Φ and fulfil (with β_N and β_L as before)

$$\mu_i := \frac{1}{2}\left[(d-i)\beta_N - i\beta_L\right].$$

1.3 Isotropic Brownian Flows

Proof: [5, (7.2) and (7.3)]. □

Due to the rotational invariance of IBFs the laws of the $V_i(x, \cdot)$ (i.e. the laws of suitable orthonormed bases) are just the uniform distribution on the unit ball i.e. the Haar measure of the topological group of rotations of $\mathbb{R}^d$. The top Lyapunov exponent μ_1 i.e. its sign crucially affects the asymptotic behaviour of the flow, as shown in [9]. It gives the drift of $|\Phi_t(x) - \Phi_t(y)|$ for very small $|x-y|$.

1.3.4 Gaussian Measures And Reproducing Kernel Hilbert Spaces

As for any Gaussian measure the law of $(M(t,x) : t \geq 0, x \in \mathbb{R}^d)$ naturally leads to a Hilbert space the so-called *reproducing kernel Hilbert space* (RKHS) or the *Cameron-Martin* space. We will give a brief introduction into this topic and quote a theorem that allows for the expansion of $M(t,x)$ in terms of complete orthonormal systems of this space. We finish with a support theorem for IBFs and closely follow [12].

Let $M(t,x)$ be the generating field of an isotropic Brownian flow. The random field $U(x) := M(1,x)$ can be canonically realized as a Gaussian measure $\tilde{\mathcal{N}}$ on the Borel σ-algebra on $C(\mathbb{R}^d : \mathbb{R}^d)$. The space $C(\mathbb{R}^d : \mathbb{R}^d)$ equipped with the topology of the uniform convergence on compacts is a locally convex topological vector space which is also metrizable via the metric

$$\tilde{\rho}(f,g) := \sum_{n=0}^{\infty} \frac{1}{2^k} \frac{\tilde{\rho}_n(f,g)}{1+\tilde{\rho}_n(f,g)} \text{ where } \tilde{\rho}_n(f,g) := \max_{i=1,\ldots,d} \sup_{|x|\leq n} |f_i(x) - g_i(x)|.$$

Remark:

Note that actually $U(x)$ is smoother than just continuous, but since we are interested in the RKHS associated with the distribution of U, [8, Lemma 3.2.2] tells us that the RKHS is the same as long as the space of the "smoother" functions is continuously and linearly embedded in the bigger space. Consider the space $C(\mathbb{R}^d \times \{1,\ldots,d\} : \mathbb{R})$ defined as

$$C(\mathbb{R}^d \times \{1,\ldots,d\} : \mathbb{R}) := \{f \colon \mathbb{R}^d \times \{1,\ldots,d\} \to \mathbb{R} : f \text{ is continuous}\},$$

where $\mathbb{R}^d \times \{1,\ldots,d\}$ is understood with the product topology.

The space $C(\mathbb{R}^d \times \{1,\ldots,d\} : \mathbb{R})$, equipped with the topology of the uniform convergence on compact sets is a locally convex set and is also metrizable via the metric

$$\rho(f,g) := \sum_{n=0}^{\infty} \frac{1}{2^n} \frac{\rho_n(f,g)}{1+\rho_n(f,g)} \quad \text{where } \rho_n(f,g) := \max_{i=1,\dots,d} \sup_{|x|\leq n} |f(x,i) - g(x,i)|.$$

It is easy to see that $C^2(\mathbb{R}^d : \mathbb{R}^d)$ and $C^2(\mathbb{R}^d \times \{1,\dots,d\} : \mathbb{R})$ can be identified through the isomorphism $\mathcal{I}$

$$\mathcal{I}\colon C^2(\mathbb{R}^d : \mathbb{R}^d) \to C^2(\mathbb{R}^d \times \{1,\dots,d\} : \mathbb{R})$$

with

$$\left(\mathbb{R}^d \ni x \xmapsto{f} (f_1(x),\dots,f_d(x))^T \in \mathbb{R}^d\right) \xmapsto{\mathcal{I}} \left(\mathbb{R}^d \times \{1,\dots,d\} \ni (x,i) \xmapsto{\mathcal{I}f} \mathcal{I}(f)(x,i) := f_i(x) \in \mathbb{R}\right).$$

The distribution of $U(x)$, as a $C(\mathbb{R}^d \times \{1,\dots,d\} : \mathbb{R})$-valued random variable is again centered Gaussian and will be denoted by $\mathcal{N}$. According to [12, Proposition D.3.1] its Cameron-Martin space $H(\mathcal{N})$ (resp. reproducing kernel Hilbert space) is the range of the covariance operator of $\mathcal{N}$ (as a bilinear functional on the dual of $C(\mathbb{R}^d : \mathbb{R}^d) \simeq C(\mathbb{R}^d \times \{1,\dots,d\} : \mathbb{R}))$ and hence generated by the symmetric positive kernel

$$k(x,i,y,j)\colon \left(\mathbb{R}^d \times \{1,\dots,d\}\right) \times \left(\mathbb{R}^d \times \{1,\dots,d\}\right) \to \mathbb{R}$$

defined via

$$k(x,i,y,j) := \int_C \delta_{(x,i)}(f)\delta_{(y,j)}(f)\mathcal{N}(df) = \int_C f(x,i)f(y,j)\mathcal{N}(df)$$
$$= \int_C f_i(x)f_j(y)\mathcal{N}(df) = \mathbb{E}\left[F^i(x,1)F^j(y,1)\right] = b_{i,j}(x-y),$$

where $\delta_{(x,i)}$ denotes the evaluation functional on $C(\mathbb{R}^d \times \{1,\dots,d\} : \mathbb{R})$. Observe that it is a continuous linear functional.

Proposition 1.3.10 (elements of the RKHS).
Let μ be a finite signed measure on $(\mathbb{R}^d \times \{1,\dots,d\}, \mathcal{B})$ having compact support. Then the function

$$h(x,i) := \int_{\mathbb{R}^d \times \{1,\dots,d\}} b_{i,j}(x-y)\mu(dy\,dj)$$

is an element of the RKHS, associated to the kernel k.
The linear subspace spanned by $\left(b_{.j}(x-\cdot) : j \in \{1,\ldots,d\},\ x \in \mathbb{R}^d\right)$ is dense in this RKHS.

Proof: [12, D.3.1] or [8]. □

Now we are ready to state the theorems mentioned before.

Theorem 1.3.11 (representation of M in terms of its RKHS).
For an isotropic Brownian field $M = M(t,x)$ let $(V_i : i \in \mathbb{N})$ be a complete orthonormal system of its associated RKHS and $(W^i : i \in \mathbb{N})$ a sequence of independent standard Brownian motions. Then we have
$$M(t,x) = \sum_{i=1}^{\infty} V_i(x) W_t^i.$$
The convergence mode is the a.s. uniform convergence of all derivatives up to order 2 on compacts, i.e. we especially have for $\mathbb{K} \subset\subset \mathbb{R}^d$ that
$$\mathbb{P}\left[\sup_{0 \leq t \leq T} \sup_{x \in \mathbb{K}} \left\|\sum_{i=1}^{n} V_i(x) W_t^i - M(t,x)\right\| \stackrel{n \to \infty}{\longrightarrow} 0\right] = 1.$$

Proof: [8, Theorem 3.5.1] and [12, Appendix D]. □

Finally we proceed to the support theorem.

Theorem 1.3.12 (support theorem for IBFs).
Let M be an isotropic Brownian field. Due to Theorem 1.3.11 this can be written as $M(t,x) = \sum_{i=1}^{\infty} V_i(x) W_t^i$ (notation as before). Assume that V_1 is four times continuously differentiable and that all derivatives up to order four are bounded. Then for $\mathbb{K} \subset\subset \mathbb{R}^d$, $T > 0$ and $\delta > 0$ there are positive numbers ϵ and C such that
$$\mathbb{P}\left[\sup_{0 \leq t \leq T} \sup_{x \in \mathbb{K}} \left\|x_{\frac{t}{C}} - \psi_t(x)\right\| < \delta\right] > \epsilon.$$
Therein $\psi = \psi_t(x)$ is the solution of the deterministic control problem
$$\begin{cases} \partial_t \psi_t(x) &= V_1(\psi_t(x)) \\ \psi_0(x) &= x \end{cases}.$$

Proof: [12, Theorem 6.2.3]. □

1.3.5 Time Reverse And Markov-Properties

The various spatial invariance properties of IBFs also imply an invariance property w.r.t. time change. Precisely if one takes $T > 0$ and considers the forward flow and the backward flow induced by an IBF on time horizon $[0, T]$ then their laws coincide. Note that this fails to be true if one replaces the deterministic T by a random time (it is in general even false for $\mathcal{F}_{s,t}$-stopping times).

Lemma 1.3.13 (time reverse for IBFs).
For arbitrary $T > 0$ we have:

$$\mathcal{L}\left[(\Phi_{s,t}(.)) : 0 \leq s \leq t \leq T)\right] = \mathcal{L}\left[(\Phi_{T-s,T-t}(.)) : 0 \leq s \leq t \leq T)\right] \tag{1.17}$$

Let $\mathcal{F}_{s,t}$ be as in Lemma 1.1.8. For any $(\mathcal{F}_{s,t} : s \in (-\infty, t])$-stopping-time τ we have

$$\mathcal{L}\left[\Phi_{\tau,t}\left(\Phi_{r,\tau}(.)\right) : r \leq \tau \mid \mathcal{F}_{\tau,t}\right] = \mathcal{L}_{\tau,t}\left[\Phi_{t+r-\tau,t}(.) : r \leq \tau\right]. \tag{1.18}$$

Proof: Due to Proposition 1.1.7 the infinitesimal generator of the backward flow is the semimartingale field $M(t,x) - \int_0^t \sum_{j=1}^d \frac{\partial b^{\cdot j}}{\partial x_j}(x-y)|_{y=x}ds$ which by Lemma 1.3.2 is nothing but $M(t,x)$. Therefore the law of the forward flow and the law of the backward flow coincide. This proves (1.17). The rest follows from this and Lemma 1.1.8. □

1.4 Isotropic Ornstein-Uhlenbeck Flows

As we have already seen the IBFs are the class of stochastic flows that generically links to Lebesgue measure for the following reasons.

1. Their laws are invariant under rigid motions just as Lebesgue measure itself. Both are characterized by this property (up to some norming constants).

2. Lebesgue measure is (modulo a multiplicative constant) the invariant measure for the RDS associated to an IBF (if one neglects the fact that it is not a probability).

The idea behind the definition of the isotropic Ornstein-Uhlenbeck flows is to put a Gaussian measure as reference measure instead of Lebesgue measure, because of its importance

1.4 Isotropic Ornstein-Uhlenbeck Flows

in probability theory. One then can hope for the RDS' to have a nice invariant probability measure (i.e. the Gaussian) and so to be able to apply the RDS theory in its full strength. Another motivation for considering the IOUFs is the natural question of letting an isotropic Brownian flow evolve in a localizing potential, i.e. for some field $U\colon \mathbb{R}^d \to \mathbb{R}$ of considering the IBF Φ^U enforced by the potential U, that is, Φ^U is generated via an SDE driven by the semimartingale field

$$V^U(t, x, \omega) = M(t, x, \omega) - \nabla U(x)t.$$

The IOUFs correspond to a quadratic potential $U = \frac{c}{2}|x|^2$. However, adding a drift typically will destroy many of the nice properties of an IBF. Nevertheless the quadratic potential given above turns out to be sufficiently nice to yield interesting properties without being totally intractable. As one obtains an Ornstein-Uhlenbeck process from a Brownian motion by subjecting it to a quadratic potential i.e. by passing from the classical Itô SDE

$$dX_t = dW_t, \quad X_0 = x_0$$

for a Brownian motion $(W_t)_{t \geq 0}$ to the classical Itô SDE

$$dX_t = -cX_t dt + dW_t, \quad X_0 = x_0$$

with the same Brownian motion $(W_t)_{t \geq 0}$ and some constant $c > 0$ one can obtain an isotropic Ornstein-Uhlenbeck flow by subjecting an IBF to the quadratic potential given above i.e. by passing from the Kunita-type SDE

$$\Phi_{s,t}(x) = \int_s^t M(du, \Phi_{s,u}(x)), \quad \Phi_{s,s}(x) = x$$

to the Kunita-type SDE

$$\phi_{s,t}(x) = \int_s^t M(du, \phi_{s,u}(x)) - c \int_s^t \phi_{s,u}(x) du, \quad \phi_{s,s}(x) = x$$

wherein $M = M(t,x)$ denotes an isotropic Brownian field (in both cases) and $c > 0$ is a constant. IOUFs have been defined first in [12] and further studied in [57], [54] and [55].

1.4.1 The Generating Field

As for IBFs we introduce the IOUFs by specifying the generator $F(t, x)$.

Definition 1.4.1 (IOUF).
Let $c > 0$ and $M(t, x, \omega)$ be an isotropic Brownian field with a C^4-isotropic covariance tensor b. We define the semimartingale field $F(t, x, \omega) := M(t, x, \omega) - cxt$ which corresponds to the choice $V(x, t) = -cxt$. An isotropic Ornstein-Uhlenbeck flow (IOUF) is defined to be the solution $\phi = \phi_{s,t}(x, \omega)$ of the Kunita-type (SDE)

$$\phi_{s,t}(x) = x + \int_s^t F(du, \phi_{s,u}(x)) = x + \int_s^t M(du, \phi_{s,u}(x)) - c \int_s^t \phi_{s,u}(x) du.$$

We call c the drift of ϕ and b the covariance tensor of ϕ.

Remark: IOUFs take values in $C^{3,\delta}$ for arbitrary $\delta > 0$. We may associate to ϕ an isotropic Brownian flow Φ corresponding to the case $c = 0$. Note that we do not want IBFs to be special IOUFs (which we ensured by $c > 0$). Care has to be taken concerning the name „isotropic" because IOUFs (or their laws respectively) are not isotropic at all: the law of an IOUF changes if one applies a translation on the state space and the origin is distinguished by being the only zero of the drift in the SDE. We stick to the name because the most important ingredient of an IOUF is an isotropic covariance tensor. Nevertheless the IOUFs retain quite a lot of the nice properties of IBFs e.g. the fact that lots of finite-dimensional projections can be given rather explicitly and turn out to be diffusions. We summarize these kind of properties in the following lemma.

Lemma 1.4.2 (some finite dimensional marginals for IOUFs).
Let ϕ be an IOUF with drift c and covariance tensor b. Then we have the following.

1. *ϕ is a Brownian Flow (i.e. it has independent increments) and its law is invariant under orthogonal transformations.*

2. *The one-point motion is an Ornstein-Uhlenbeck diffusion with generator*

$$\mathcal{L} := -c \sum_{i=1}^d x_i \frac{\partial}{\partial x_i} + \frac{1}{2} \sum_{i=1}^d \frac{\partial^2}{\partial x_i^2} = -c \sum_{i=1}^d x_i \frac{\partial}{\partial x_i} + \frac{1}{2} \Delta. \quad (1.19)$$

3. The difference process $(\phi_t(x) - \phi_t(y) : t \in \mathbb{R}_+)$ is a diffusion with generator

$$\mathcal{L}_d := -c \sum_{i=1}^{d} x_i \frac{\partial}{\partial x_i} + \sum_{i,j=1}^{d} (\delta_{ij} - b_{ij}(x)) \frac{\partial^2}{\partial x_i \partial x_j}.$$

4. The spatial derivative $D_x \phi$ solves the SDE

$$D_x \phi_{0,t}(\cdot) = \mathrm{id}_{\mathbb{R}^d} + \int_0^t D_{\phi_{0,u}(\cdot)} F(du, \phi_{0,u}(\cdot)) D_x \phi_{0,u}(\cdot) \qquad (1.20)$$

which reads in components for $1 \leq i, j \leq d$

$$\partial_j \phi_t^i(x) = \delta_{ij} + \int_0^t \sum_k \partial_j \phi_s^k(x) \, \partial_k M^i(ds, x_s) - c \int_0^t \partial_j \phi_t^i(x) \, ds.$$

5. The distance process $(|\phi_t(x) - \phi_t(y)| : t \in \mathbb{R}_+)$ is a diffusion with generator

$$\mathcal{A} := (1 - B_L(r)) \frac{d^2}{dr^2} + \left((d-1) \frac{1 - B_N(r)}{r} - cr \right) \frac{d}{dr}. \qquad (1.21)$$

Proof: [12, Proposition 7.1.1] or [57, Proposition 2.2]. [12] also includes a recurrence-transience classification of the distance process. See also [54]. □

The next lemma will be used to control the moments of the two-point distance in terms of a SDE.

Lemma 1.4.3 (Two-Point-Control IOUF version).
Let $(\phi_{s,t} : 0 \leq s \leq t < \infty)$ be an IOUF. There are constants $\lambda > 0$ and $\bar{\sigma} > 0$ such that for $x, y \in \mathbb{R}^d$ there is a standard Brownian motion $(W_t)_{t \geq 0}$ such that we have for $(x_t, y_t) := (\phi_t(x), \phi_t(y))$ the following.

1. We have a.s. for all $t \geq 0$ that

$$|x_t - y_t| \leq |x - y| e^{\bar{\sigma} \sup_{0 \leq s \leq t} W_s + \lambda t}.$$

2. We have for each $x, y \in \mathbb{R}^d$, $T > 0$ and $q \geq 1$ that

$$\mathbb{E}\left[\sup_{0 \leq t \leq T} |x_t - y_t|^q \right]^{1/q} \leq 2|x - y| e^{(\lambda + \frac{1}{2} q \bar{\sigma}^2) T}.$$

Proof: [12, Corollary 7.1.1] or [57, Corollary 2.3] and [52, Condition (H)]. □

Observe that this lemma holds for a very general class of stochastic flows. In addition IOUFs also retain some of the global properties of IBFs which we summarize in the next lemma.

Lemma 1.4.4 (general properties of IOUFs).
Let $(\phi_{s,t} : 0 \leq s, t < \infty)$ be an IOUF. Then it is a Brownian flow that satisfies the following.

1. *It is temporally homogenous i.e. for $C > 0$ the laws of $(\phi_{s,t} : 0 \leq s, t < \infty)$ and $(\phi_{s+C,t+C} : 0 \leq s, t < \infty)$ coincide.*

2. *It is rotation invariant i.e. for any orthogonal matrix O we have that $(O\phi_{s,t}(\cdot) : 0 \leq s, t < \infty)$ and $(\phi_{s,t}(O\cdot) : 0 \leq s, t < \infty)$ coincide in law.*

3. *It has independent increments i.e. for $0 < t_1 < \ldots < t_n < \infty$ we have that the random mappings $(\phi_{0,t_1}, \phi_{t_1,t_2}, \ldots, \phi_{t_{n-1},t_n})$ are independent.*

4. *The spatial derivative at x is spatially homogenous, i.e. for $x, y \in \mathbb{R}^d$ the laws of $(D\phi_t(x) : t \geq 0)$ and $(D\phi_t(y) : t \geq 0)$ coincide.*

Proof: 1.-3. follow directly from the properties of the driving field and 4. follows from (1.20). □

1.4.2 Invariant Measures

For the application of Theorem 1.2.9 requires the existence of an invariant probability we will pay some attention to the question of existence of invariant measures for IOUFs although the results are well known.

Proposition 1.4.5 (invariant measure for the one-point motion).
Let ϕ be an IOUF with covariance tensor b and drift c. Then we have that the measure $\mu := \mathcal{N}(0, \frac{2}{c})^{\otimes d}$ with density $x = (x_1, \ldots, x_d) \mapsto \left(\frac{c}{\pi}\right)^{\frac{d}{2}} e^{-c|x|^2}$ is an invariant measure in the sense of Markov processes for the one-point motion of an IOUF i.e. we have for all $t \geq 0$ that

$$\mu = P_t \mu$$

where P_t denotes the semigroup of the one-point motion of an IOUF (cf. [12, Remark 9.1]).

1.4 Isotropic Ornstein-Uhlenbeck Flows

Proof: The proof relies on the fact that all the measures under consideration are locally finite and hence it is sufficient to check that the generator $\mathcal{L}$ of the one-point motion (given by (1.19) for different values of c) and the measure μ (suspected to be invariant) satisfy for any function $f \in C_0^\infty$ (cf. [12, Proposition B.2] and the references therein) the equality $\int_{\mathbb{R}^d} \mathcal{L}f(x)\mu(dx) = 0$. This can be checked by using integration by parts in the following manner (observe that we write $\sum_i$ for $\sum_{i=1}^d$ i.e. we omit the summation set when it is clear from the context).

$$\begin{aligned}
\int_{\mathbb{R}^d} \mathcal{L}f(x)\mu(dx) &= \left(\frac{c}{\pi}\right)^{d/2} \int_{\mathbb{R}^d} \left[-c\sum_i x_i \partial_i f(x) + \frac{1}{2}\sum_i \partial_i \partial_i f(x)\right] e^{-c|x|^2} dx \\
&= \left(\frac{c}{\pi}\right)^{d/2} \left(\int_{\mathbb{R}^d} \frac{1}{2}\sum_i e^{-c|x|^2} \partial_i \partial_i f(x) dx - c\int_{\mathbb{R}^d} \sum_i x_i e^{-c|x|^2} \partial_i f(x) dx\right) \\
&= \left(\frac{c}{\pi}\right)^{d/2} \left(-\int_{\mathbb{R}^d} \frac{1}{2}\sum_i \left(\partial_i e^{-c|x|^2}\right) \partial_i f(x) dx - c\int_{\mathbb{R}^d} \sum_i x_i e^{-c|x|^2} \partial_i f(x) dx\right) \\
&= \left(\frac{c}{\pi}\right)^{d/2} \left(\int_{\mathbb{R}^d} \frac{1}{2}\sum_i 2cx_i e^{-c|x|^2} \partial_i f(x) dx - c\int_{\mathbb{R}^d} \sum_i x_i e^{-c|x|^2} \partial_i f(x) dx\right) = 0.
\end{aligned}$$

$\square$

The above shows that the one-point motion of an IOUF has an invariant measure in the sense of Markov processes and to be accurate we have to check that the same measure is an invariant measure in the sense of random dynamical systems (to be precise we have to take $\mu \otimes \tilde{\mathbb{P}}$).

1.4.3 Lyapunov Exponents

The Lyapunov spectrum for an IOUF can be computed from the Lyapunov spectrum of the IBF that one gets by neglecting the drift c of the IOUF. This has been carried out in [57] and we only state the result.

Lemma 1.4.6 (Lyapunov exponents for IOUFs).
Let $(\phi_{s,t} : 0 \leq s, t < \infty)$ be an IOUF with covariance tensor b and drift c. Then it has d different Lyapunov exponents in the sense of Theorem 1.2.9, which are given by

$$\lambda_i := (d-i)\frac{\beta_N}{2} - i\frac{\beta_L}{2} - c.$$

In particular they all have simple multiplicities.

Proof: [57, Proposition 2.5]. □

1.5 Repulsive Isotropic Flows

We define the notion of repulsive isotropic flows (RIF) to cover the case of an IOUF SDE with drift $c < 0$ where this can be done for free. Although it seems possible to derive quite a lot of properties of RIFs using standard methods we limit ourselves to the extension of new results on IOUFs to the case $c < 0$. Since the RIF are not of central importance for the following we keep the following as short as possible.

Definition 1.5.1 (repulsive isotropic flow).
Let $c < 0$ and $M(t, x, \omega)$ be an isotropic Brownian field with a C^4-isotropic covariance tensor b. We define the semimartingale field $F(t, x, \omega) := M(t, x, \omega) - cxt$ which (again) corresponds to the choice $V(t, x) = -cxt$. A repulsive isotropic flow (RIF) is defined to be the solution $\phi = \phi_{s,t}(x, \omega)$ of the Kunita-type (SDE)

$$\phi_{s,t}(x) = x + \int_s^t F(du, \phi_{s,u}(x)) = x + \int_s^t M(du, \phi_{s,u}(x)) - c\int_s^t \phi_{s,u}(x)du.$$

We call c the drift of ϕ and b the covariance tensor of ϕ.

Remark: Unifying the notation we introduce a name for IOUFs IBFs and RIFs together and speak *isotropic flows*. Since all the argueing about the names above applies we suggest that a name like „rotation invariant flows" should be prefered. Nevertheless we call IBFs, IOUFs and RIFs *isotropic flows* although it is not clear how the invariance under rotation links exactly to the driving SDEs we consider.

Theorem 1.5.2 (general properties of RIFs).
Let ϕ be an RIF with drift $c < 0$ and covariance tensor b. Then we have the following.

1. *All the assertions of Lemma 1.4.2 (except the claim that the generator of the one-point motion belongs to an asymptotically stationary Ornstein-Uhlenbeck process) hold.*

2. *For any $x \in \mathbb{R}^d$ we have that $\phi_t(x) - x$ is distributed as $\mathcal{N}(0, \frac{1}{2c}(1 - e^{-2ct}))$.*

3. *The assertions of Lemma 1.4.3 hold.*

Proof: For 1. it is enough to observe that the proof in [12] is valid for arbitrary $c \neq 0$. Note that one can solve the SDE for the one-point motion and obtain 2. from this just as in the IOUF case. The difference is that 2. excludes the possibility that the one-point motion might become stationary because of the increasing variance. 3. follows from [52, Lemma 4.1]. □

Chapter 2

Spatial Regularity

In this chapter we prove a result concerning the macroscopic bevahiour of RIFs, IBFs and IOUFs on a fixed time horizon. We show that their stochasticity is negligible far away from the origin if compared to the distance to the origin. The asymptotic spatial behavior of stochastic flows and their derivatives (of any order!) has been studied in [20] in a very general setting but nevertheless with a slightly different scope. Since we only consider the special cases of IOUFs and IBFs we can give a more specific result (which in this simplicity cannot be expected to hold in the case of a general drift). It is somewhat clear that the obtained results are not sharp but since we need them (for IOUFs) in Chapter 5 precisely in the form given below we do not care about generalizations.

2.1 Preliminaries

The proof of the main theorem depends crucially on the *chaining technique*. We will use the formulation from [10], where also a proof can be found.

Let $(\mathbb{X}, \rho)$ be a compact metric space and $\phi \colon \mathbb{X} \to \mathbb{R}_+$ be a random continuous function i.e. a random variable taking values in the set of continuous functions from $\mathbb{X}$ to $\mathbb{R}_+$. Given a sequence of positive real numbers $(\delta_i)_{i \geq 0}$, such that $\sum_{i=0}^{\infty} \delta_i < \infty$ we determine a sequence $(\chi_i)_{i=0}^{\infty}$ of discretizations (skeletons) of $\mathbb{X}$, with the property that for all $x \in \mathbb{X}$ there is a point $x_i \in \chi_i$, such that $\rho(x, x_i) \leq \delta_i$. Assume that $\chi_0 = \{x_0\}$, with $\rho(x, x_0) \leq \delta_0$ for all $x \in \mathbb{X}$.

Proposition 2.1.1 (Chaining).
Let $\phi\colon \mathbb{X} \to \mathbb{R}_+$ be an almost surely continuous random function with $(\delta_i)_{i\geq 0}$ and $(\chi_i)_{i=0}^{\infty}$ as above. For arbitrary positive $\epsilon, z \geq 0$ and an arbitrary sequence of positive reals $(\epsilon_i)_{i\geq 0}$ such that $\epsilon + \sum_{i=0}^{\infty} \epsilon_i = 1$ we have

$$\mathbb{P}\Big(\sup_{x\in\mathbb{X}} \phi(x) > z\Big) \leq \mathbb{P}\big(\phi(x_0) > \epsilon z\big) + \sum_{i=0}^{\infty} |\chi_{i+1}| \sup_{\rho(x,y)\leq \delta_i} \mathbb{P}\big(|\phi(x) - \phi(y)| > \epsilon_i z\big).$$

Proof: [10, Lemma 4.1]. □

The following lemma simply states some well known facts about the running maximum of a standard Brownian motion as well as a common estimate for the Gaussian tail. We will use them frequently and therefore state them explicitly.

Lemma 2.1.2 (Gaussian tails).
Let $(W_t)_{t\geq 0}$ be a standard Brownian motion, and let $W_t^\star := \sup_{s\leq t} W_s$ be its running maximum. The distribution of $W_t^\star$ has density

$$\mathbb{1}_{[0,\infty)}(x) \sqrt{\frac{2}{\pi t}} e^{-\frac{x^2}{2t}}$$

with respect to the Lebesgue measure. Moreover for arbitrary $K > 0$ the following bounds hold.

$$\mathbb{P}(W_t \geq K) \leq \frac{1}{K}\sqrt{\frac{t}{2\pi}} e^{-\frac{K^2}{2t}} \quad \text{and} \quad \mathbb{P}(W_t^\star \geq K) \leq \frac{1}{K}\sqrt{\frac{2t}{\pi}} e^{-\frac{K^2}{2t}}.$$

Proof: One only has to observe that the running maximum and the modulus of a Brownian motion have the same one-dimensional distributions. □

2.2 Spatial Regularity Lemma

The one-point motion $\phi_t(x)$ of an IOUF is as already stated an Ornstein-Uhlenbeck process and so if x is far away from the origin $|\phi_t(x)|$ will decrease roughly as $|x|e^{-ct}$ (which is the expected decrease) because the variance of $|\phi_t(x)|$ is rather negligible for large $|x|$. We may expect the IOUFs unitstep-discretization $\phi = \phi_{0,1}$ to look like e^{-c} times the identity

2.2 Spatial Regularity Lemma

on a large scale. For a fixed x this is quite obvious but care has to be taken about the fact that we are dealing with infinitely many random variables. The next lemma states that this is no problem at all. The same is true for IBFs and RIFs (with the appropriate modifications).

Lemma 2.2.1 (spatial asymptotics for IBFs and IOUFs).
Let $(\phi_{s,t} : 0 \leq s \leq t < \infty)$ be an IOUF as in Definition 1.4.1 or a RIF as in Definition 1.5.1. Then we have for any $t > 0$ a.s. the following ($c \neq 0$ denotes the drift in both cases).

1.
$$\lim_{R \to \infty} \sup_{|x| \geq R} \frac{|\phi_t(x) - e^{-ct}x|}{|x|} = 0. \tag{2.1}$$

2.
$$\lim_{R \to \infty} \sup_{|x| \geq R} \frac{|\phi_t(x)|}{|x|} = e^{-ct}.$$

Let $(\Phi_{s,t} : 0 \leq s \leq t < \infty)$ be an IBF as in Definition 1.3.3. Then we have for any $t > 0$ a.s. the following.

3.
$$\lim_{R \to \infty} \sup_{|x| \geq R} \frac{|\Phi_t(x) - x|}{|x|} = 0. \tag{2.2}$$

4.
$$\lim_{R \to \infty} \sup_{|x| \geq R} \frac{|\Phi_t(x)|}{|x|} = 1.$$

Proof: Observe that in both cases the second formula is an easy consequence of the first one, so we will only have to prove (2.1) and (2.2). Therefore it is sufficient to show

$$\lim_{R \to \infty} \sup_{R \leq |x| \leq R+1} \frac{|\phi_t(x) - e^{-ct}x|}{|x|} = 0 \text{ and } \lim_{R \to \infty} \sup_{R \leq |x| \leq R+1} \frac{|\Phi_t(x) - x|}{|x|} = 0 \text{ a.s..}$$

Let $A_R := \{x \in \mathbb{R}^d : R \leq |x| \leq R+1\}$. We want to apply the chaining technique (see Proposition 2.1.1) and so we first observe that there is a constant c_1 such that A_R can be covered by $c_1 R^d 3^{j-1}$ balls of radius $\delta_j = 3^{-j}$ if $j \geq 1$. Let χ_j consist of the centers of these balls and $\epsilon_j = 2^{-j-2}$ as well as $\epsilon = \frac{1}{2}$. Fixing $\tilde{\epsilon} > 0$ and an arbitrary $x_0 \in A_R$ (with

$\delta_0 := 2(R+1)$) we conclude

$$\mathbb{P}\left[\sup_{x \in A_R} \frac{|\phi_t(x) - e^{-ct}x|}{|x|} > \tilde{\epsilon}\right] \leq \mathbb{P}\left[\sup_{x \in A_R} |\phi_t(x) - e^{-ct}x| > \tilde{\epsilon}R\right] \leq \mathbb{P}\left[|\phi_t(x_0) - e^{-ct}x_0| > \frac{\tilde{\epsilon}R}{2}\right]$$
$$+ \sum_{j=0}^{\infty} c_1 R^d 3^j \sup_{|x-y| \leq 3^{-j}} \mathbb{P}\left[||\phi_t(x) - e^{-ct}x| - |\phi_t(y) - e^{-ct}y|| > 2^{-j-2}\tilde{\epsilon}R\right]$$

as well as

$$\mathbb{P}\left[\sup_{x \in A_R} \frac{|\Phi_t(x) - x|}{|x|} > \tilde{\epsilon}\right] \leq \mathbb{P}\left[\sup_{x \in A_R} |\Phi_t(x) - x| > \tilde{\epsilon}R\right] \leq \mathbb{P}\left[|\Phi_t(x_0) - x_0| > \frac{\tilde{\epsilon}R}{2}\right]$$
$$+ \sum_{j=0}^{\infty} c_1 R^d 3^j \sup_{|x-y| \leq 3^{-j}} \mathbb{P}\left[||\Phi_t(x) - x| - |\Phi_t(y) - y|| > 2^{-j-2}\tilde{\epsilon}R\right].$$

Using some standard estimates for the normal distribution (as stated in Lemma 2.1.2 for a Brownian motion) we conclude that

$$\mathbb{P}\left[|\phi_t(x_0) - e^{-tc}x_0| > \frac{\tilde{\epsilon}R}{2}\right] \leq d\mathbb{P}\left[|\phi_t^1(x_0) - e^{-tc}x_0^1| > \frac{\tilde{\epsilon}R}{2d}\right]$$
$$= 2d\mathcal{N}\left(0, \frac{1}{2c}(1 - e^{-2ct})\right)\left(\left(\frac{\tilde{\epsilon}R}{2d}, \infty\right)\right) \leq \frac{2d^2}{\tilde{\epsilon}R}\sqrt{\frac{1-e^{-2ct}}{\pi c}}e^{-\frac{1}{2}(\frac{\tilde{\epsilon}R}{2d})^2\frac{2c}{1-e^{-2ct}}} \leq c_2 e^{-c_3 R^2}$$

as well as

$$\mathbb{P}\left[|\Phi_t(x_0) - x_0| > \frac{\tilde{\epsilon}R}{2}\right] \leq d\mathbb{P}\left[|\Phi_t^1(x_0) - x_0^1| > \frac{\tilde{\epsilon}R}{2d}\right]$$
$$= 2d\mathcal{N}(0,t)\left(\left(\frac{\tilde{\epsilon}R}{2d}, \infty\right)\right) \leq \frac{2d^2}{\tilde{\epsilon}R}\sqrt{\frac{2t}{\pi}}e^{-\frac{1}{2t}(\frac{\tilde{\epsilon}R}{2d})^2} \leq c_2 e^{-c_3 R^2}$$

where c_2 and c_3 are constants that only depend on c, d, $\tilde{\epsilon}$ and t (but not on R). W.l.o.g. we now assume that $R \geq 16e^{-ct}\tilde{\epsilon}^{-1} \vee 8\tilde{\epsilon}^{-1}$. Using Lemmas 1.3.8, 1.4.3 and 2.1.2 as well as

2.2 Spatial Regularity Lemma

Theorem 1.5.2 and the reversed triangle inequality we get for $|x-y| \leq 3^{-j}$ that

$$\mathbb{P}\left[||\phi_t(x) - e^{-ct}x| - |\phi_t(y) - e^{-ct}y|| > 2^{-j-2}\tilde{\epsilon}R\right]$$
$$\leq \mathbb{P}\left[||\phi_t(x) - e^{-ct}x| - |\phi_t(y) - e^{-ct}y|| > 2^{-j-2}\tilde{\epsilon}R3^j|x-y|\right]$$
$$\leq \mathbb{P}\left[|\phi_t(x) - e^{-ct}x - \phi_t(y) + e^{-ct}y| > 2^{-j-2}3^j\tilde{\epsilon}R|x-y|\right]$$
$$\leq \mathbb{P}\left[|\phi_t(x) - \phi_t(y)| > 2^{-j-3}3^j\tilde{\epsilon}R|x-y|\right] + \mathbb{P}\left[e^{-ct}|x-y| > 2^{-j-3}3^j\tilde{\epsilon}R|x-y|\right]$$
$$\leq \mathbb{P}\left[W_t^* \geq \frac{\log(2^{-3-j}3^j\tilde{\epsilon}R) - \lambda t}{\bar{\sigma}}\right]$$

and similarly

$$\mathbb{P}\left[||\Phi_t(x) - x| - |\Phi_t(y) - y|| > 2^{-j-2}\tilde{\epsilon}R\right]$$
$$\leq \mathbb{P}\left[|\Phi_t(x) - \Phi_t(y)| > 2^{-j-3}3^j\tilde{\epsilon}R|x-y|\right] + \mathbb{P}\left[|x-y| > 2^{-j-3}3^j\tilde{\epsilon}R|x-y|\right]$$
$$\leq \mathbb{P}\left[W_t^* \geq \frac{\log(2^{-3-j}3^j\tilde{\epsilon}R) - \lambda t}{\bar{\sigma}}\right]$$

where the Brownian motions and the constants λ and $\bar{\sigma}$ come from Lemmas 1.4.3 and 1.3.8 as well as Theorem 1.5.2 respectively. Of course assuming $R \geq e^{\frac{8}{\tilde{\epsilon}}(\lambda t+1)}$ we have that

$$\mathbb{P}\left[W_t^* \geq \frac{\log(2^{-3-j}3^j\tilde{\epsilon}R) - \lambda t}{\bar{\sigma}}\right] \leq \frac{\bar{\sigma}}{\log(2^{-3-j}3^j\tilde{\epsilon}R) - \lambda t}\sqrt{\frac{2t}{\pi}}e^{-\frac{(\log(2^{-3-j}3^j\tilde{\epsilon}R) - \lambda t)^2}{2t\bar{\sigma}^2}}$$
$$\leq c_4(2^{-j-3}3^j\tilde{\epsilon}R)^{-\frac{\log(2^{-3-j}3^j\tilde{\epsilon}R) - 2\lambda t}{2t\bar{\sigma}^2}}$$

where the constant c_4 depends only on $\bar{\sigma}$, λ and t. Combining all the above estimates we conclude that for $\tilde{\epsilon} > 0$ we have that both probabilities $\mathbb{P}\left[\sup_{x \in A_R} \frac{|\phi_t(x) - e^{-ct}x|}{|x|} > \tilde{\epsilon}\right]$ and $\mathbb{P}\left[\sup_{x \in A_R} \frac{|\Phi_t(x) - x|}{|x|} > \tilde{\epsilon}\right]$ can be bounded from above by

$$c_2 e^{-c_3 R^2} + \sum_{j=0}^{\infty} c_1 R^d 3^j c_4 (2^{-j-3}3^j\tilde{\epsilon}R)^{-\frac{\log(2^{-3-j}3^j\tilde{\epsilon}R) - 2\lambda t}{2t\bar{\sigma}^2}}$$

which can be summed up over R. Thus we get the desired conclusion via an application of the first Borel-Cantelli Lemma. □

Chapter 3

Densities For The Finite-Point Motions

As we have already seen, the one-point motion of an IBF (IOUF) is a standard Brownian motion (Ornstein-Uhlenbeck process) and so of course possesses a C^∞-density. This chapter is devoted to the question whether this is true for finite-dimensional diffusion of the considered flows. The question of existence and smoothness of densities (w.r.t. Lebesgue measure) for the solution of SDEs is usually investigated with techniques coming from Malliavin calculus (this is the target the Malliavin calculus was originally developed for). Virtually all the literature assumes that the driving vector fields of some Stratonovich SDE are smooth and that the Lie algebra generated by them spans $\mathbb{R}^d$ at every point in $\mathbb{R}^d$ (Hörmander's strong hypothesis) or at least at the initial value of the SDE (Hörmander's weak hypothesis). Under this assumption it is shown that the solution possesses a smooth (i.e. C^∞) density with bounded derivatives. This is known as Hörmander's theorem. See [18] for the original analytic proof and e.g. [38], [16], [41], [6], [59], [40] and [19] for proofs using the Malliavin calculus and related works. The reason why it is hard to find C^k-versions of Hörmander's Theorem in the literature seems to be that Hörmander's condition is not well-defined if the vector fields are not smooth (the Lie bracket of two vector fields is a vector field if their second partial derivatives commute). Nevertheless the contents of this section seems to be well-known (but unstated in the literature) except for the (trivial) application to IBFs and IOUFs.

3.1 Existence Of Densities

We start with the following key lemma.

Lemma 3.1.1 (Norris' Lemma).
Let X be a $\mathbb{R}^d$-valued random variable and $n \geq d+1$. Assume that for all multi-indices α with $|\alpha| \leq n$ and for any test function $f \in C_b^n(\mathbb{R}^d)$ we have

$$\mathbb{E}\left[D^\alpha f(X)\right] \leq K\|f\|_\infty \tag{3.1}$$

wherein K may depend on n but not on f. Then the distribution of X is absolutely continuous w.r.t. Lebesgue measure on $\mathbb{R}^d$ and the density is $n-d-1$ times continuously differentiable.

Proof: [40, Theorem 0.1]. □

Theorem 3.1.2 (existence result for densities).
Let $k \geq 1$ be odd and let $V_0, \ldots, V_n$ denote C^{k+1} vector fields on $\mathbb{R}^d$ with bounded derivatives up to order $k+1$. Suppose that for $x \in \mathbb{R}^d$ the vectors

$$\left\{V_{i_1^1}(x), \left[V_{i_1^2}, V_{i_2^2}\right], \ldots, \left[V_{i_1^k}, \left[V_{i_2^k}, \left[\ldots, V_{i_k^k}\right]\ldots\right]\right] : i_l^j \in \{1, \ldots, n\}, 1 \leq j \leq l, 1 \leq l \leq k\right\} \tag{3.2}$$

i.e. the Lie brackets generated by $\{V_1, \ldots, V_n\}$ of order at most k span $\mathbb{R}^d$ at x (one has to require the vector field to be C^{k+1} to ensure that the Lie-brackets up to order k are well-defined). Let X_t be the solution to

$$X_t = x + \int_0^t V_0(X_s)ds + \sum_{i=1}^n V_i(X_s) \circ dW_s^{(i)}$$

for a n-dimensional Brownian motion $(W_t)_{t \geq 0}$. Then the distribution of X_t is absolutely continuous w.r.t. Lebesgue measure on $\mathbb{R}^d$ and if $k+1 \geq 2d+2$ the density is $\frac{k+1}{2} - d - 1$ times continuously differentiable.

Proof: The case $k = 1$ also follows directly from [6, Theorem 4.9] so we can focus on the required smoothness. We limit ourselves to a version of the proof which is somewhat sketchy. Consider the Malliavin covariance matrix

$$\sigma_t := Z_t^{-1} \left[\int_0^t Z_s V(X_s) V^*(X_s) ds \right] \left[Z_t^{-1} \right]^*$$

wherein V is the matrix $V = (V_1, \ldots, V_n)$ and Z_t is the solution to the SDE

$$Z_s = E_n - \int_0^s Z_u DV_0(X_u) du - \sum_{i=1}^n \int_0^s Z_u DV_i(X_u) d \circ W_u^{(i)}.$$

Then [6, Theorem 4.9] tells us that we have to verify that σ_t is a.s. invertible for any $t > 0$. The proof of [6, Theorem 6.4] shows this by proving that the vector fields apearing in (3.2) lie in the range of σ_t for any $t > 0$. [6, Section 6.2] shows even more namely that $\sigma_t \in \bigcap_{1 \leq q < \infty} L^q(\mathbb{P})$. According to Lemma 3.1.1 we now have to show that (3.1) holds for $n = \frac{k+1}{2}$. This is done in some detail in the proof of [6, Theorem 4.10]. □

At this point we are ready to derive some applications to IBFs and IOUFs as corollaries to Theorem 3.1.2.

Corollary 3.1.3 (the two-point motion).
Let (x_t, y_t) be the two-point motion of a two-dimensional IBF or IOUF with a C_b^6 covariance tensor b. Then (x_t, y_t) has a continuous density w.r.t. Lebesgue measure on $\mathbb{R}^4$.

Proof: This is a trivial combination of Theorem 3.1.2 and Lemma 1.3.7 which in fact yields ellipticity and not only hypoellipticity. □

3.2 A-Little-Positivity Of Densities

This section is devoted to the question what can be said about the positivity of the density of the 2-point-motion of an IBF or IOUF in $\mathbb{R}^d$ which exists as we already saw in Theorem 3.1.2. We will make use of the following lemma.

Lemma 3.2.1 (support theorem for elliptic diffusions).
There is $p = p(C_4, C_5, C_6, \delta, t) > 0$, such that for any $\mathbb{R}^d$-valued semimartingale S_t with martingale part M_t, satisfying for $t \leq \tau := \inf\{s : \|S_s\| \geq \delta\}$ the following

1. $S_0 = 0$,

2. *The part of locally bounded variation is Lipschitz-continuous with constant C_6,*

3. $C_4 \|z\|^2 \leq z^T \frac{d\langle M \rangle}{dt} z \leq C_5 \|z\|^2$,

we have the estimate $\mathbb{P}[\tau > t] > p$. *p can be chosen to be continuous in all variables, decreasing in C_5, C_6 and t as well as increasing in δ and $0 < C_4 \leq C_5$.*

Proof: [53, Lemma 2.4]. □

Now we come to the definition of the notion of positivity that allows for the use of rather primitive methods.

Definition 3.2.2 (a little positive).
We say that a function f on $\mathbb{R}^d$ is a little positive if there exists no open non-empty $U \subset \mathbb{R}^d$ such that the restriction of f on U vanishes identically.

Theorem 3.2.3 (little positivity theorem).
Let $(x_s, y_s)_{s \geq 0}$ be the two-point motion of a d-dimensional IBF or IOUF such that it admits a density w.r.t. Lebesgue measure on $\mathbb{R}^{2d}$. Then this density is a little positive.

Proof: Assume that this is wrong and let w.l.o.g. U be the interior of $K_\epsilon(0) \times K_\epsilon(0)$ for some $\epsilon > 0$. Let $p_s(x, y, x', y')$ denote the transition density of (x_s, y_s) at (x', y') such that $p_s(x, y, \cdot, \cdot)$ vanishes on U.
Step 1: Assume now that $\overline{x, x'}$ and $\overline{y, y'}$ are disjoint and define the $2d$-dimensional semi-martingale

$$S_t := \begin{pmatrix} x_{st} \\ y_{st} \end{pmatrix} - \left[ts \begin{pmatrix} x' \\ y' \end{pmatrix} + (1-t)s \begin{pmatrix} x \\ y \end{pmatrix} \right]$$

Then Lemma 3.2.1 tells us that

$$\mathbb{P}\left[\begin{pmatrix} x_s \\ y_s \end{pmatrix} \in U \right] \geq \mathbb{P}[|x_1 - x'| < \epsilon; |y_1 - y'| < \epsilon] \geq p(C_4, C_5, s \left\| \begin{pmatrix} x - x' \\ y - y' \end{pmatrix} \right\|, \epsilon, 1) > 0$$

for some constants C_4 and C_5 depending on $\min_{t \in [0,1]} |tsx' + (1-t)sx - tsy' - (1-t)sy|$.
This contradicts the definition of U.
Step 2: For the general case $\overline{x, x'} \cap \overline{y, y'} \neq \emptyset$ one just has to divide the action into two time steps satisfying the assumptions of Step 1. □

3.3 POSITIVITY OF DENSITIES

Remark: It is fairly natural to expect that this also holds for any finite-point diffusions of IBFs and IOUFs. But since the property "a little positive" turns out to be not very satisfying in applications we will not give more details on that (observe that not even the flow property prevents the density to be a little positive although the diffusion will never arrive at $\{x = y\}$). The next section gives a sufficient condition for the density to be positive everywhere but on this diagonal.

3.3 Positivity Of Densities

Let us again focus on the case of the two-point motion (x_t, y_t) of an d-dimensional IBF or IOUF ($x \neq y$). The homeomorphic properties of the flow do not allow for $x_t = y_t$ to hold at any time except on a null set (remember that we decided to modify the flow in a way such that $x_t = y_t$ is impossible). One might expect the process (x_t, y_t) to posesses a density on the set
$\mathbb{R}^{2d}_\times := \mathbb{R}^{2d} \setminus \{(x,y) \in \mathbb{R}^{2d} : x = y\}$. If b is smooth this is in fact true as we shall see in the following.

Theorem 3.3.1 (sufficient condition for positive density).
The two-point-motion (x_t, y_t) interpreted as a diffusion on $\mathbb{R}^{2d}_\times$ posseses a strictly positive C^∞-density on $\mathbb{R}^{2d}_\times$ provided that the covariance tensor b is C^∞ and all its derivatives are bounded.

Proof: We restrict ourselves to $t = 1$ by scaling. First observe that our smoothness assumptions on b now allow for the use of Hörmander's Theorem [18]. See [41] for details and stochastic interpretations. Since we already observed that the process satisfies the SDE (1.14) and since Lemma 1.3.7 ensures that Hörmander's strong condition is satisfied we can conclude that on $\mathbb{R}^{2d}_\times$ a C^∞-density exists. We now have to show that it is strictly positive there. We want to apply the results of [29], so we have to consider the following control problem.

$$dz_t(h) = \bar{b}(z_t(h))h_t dt$$

Therein h is a square-integrable, $\mathbb{R}^{2d}$-valued control function (in fact chosen to be continuously differentiable). z_t is a 2d-dimensional process to be thought of as a deterministic version of the two-point motion. Fix $(x, y) \in \mathbb{R}^{2d}_\times$. In order to show that (x_1, y_1) has positive transition density for any $x^{(1)}, y^{(1)} \in \mathbb{R}^{2d}_\times$ it is enough to establish the following Bismut Condition (see [7]).

Condition 3.3.2 (Bismut's condition).
For any $(x, y) = z \in \mathbb{R}^{2d}_\times, (x^{(1)}, y^{(1)}) \in \mathbb{R}^{2d}_\times$ there is an $h \in L^2$ such that

$$z_1(h) = (x^{(1)}, y^{(1)}) \tag{3.3}$$

and such that $h \mapsto (z_1(h))$ is a submersion in h. (we identify $\mathbb{R}^{2d}$ and $\mathbb{R}^d \times \mathbb{R}^d$ in the obvious way).

Proof.: Step 1: Let us assume first that $\overline{x, x^{(1)}}$ and $\overline{y, y^{(1)}}$ are disjoint and that each of them consists at least of two points. ($\overline{x,y}$ denoting the convex hull of x and y.) We construct a control satisfying (3.3) such that the stream lines of z_t are exactly $\overline{x, x^{(1)}} \cup \overline{y, y^{(1)}}$. This ensures that $\bar{b}(z_t(h))$ is regular and its determinant is bounded away from zero for all t. The simplest way to obtain the desired streamlines is to ensure $\bar{b}(z_t(h)) h_t \equiv \left(z^{(1)} - \begin{pmatrix} x \\ y \end{pmatrix} \right)$. We may hope to achieve this by setting

$$h_0 := \bar{b}\left(\begin{pmatrix} x \\ y \end{pmatrix}\right)^{-1} \left(z^{(1)} - \begin{pmatrix} x \\ y \end{pmatrix} \right)$$

as well as

$$0 = \frac{d}{dt}\left[\bar{b}(z_t(h)) h_t\right]$$

$$\Leftrightarrow \forall i = 1, \ldots, 2d : \sum_{k=1}^{2d} \left[\frac{d}{dt} \bar{b}_{i,k}(z_t(h))\right] h_{k,t} + \bar{b}_{i,k}(z_t(h)) \frac{dh_{k,t}}{dt} = 0$$

$$\Leftrightarrow \forall i = 1, \ldots, 2d : \sum_{k,l=1}^{2d} \partial_l \bar{b}_{i,k}(z_t(h)) \frac{dz_{l,t}(h)}{dt} h_{k,t} + \sum_{k=1}^{2d} \bar{b}_{i,k}(z_t(h)) \frac{dh_{k,t}}{dt}$$

$$\Leftrightarrow \bar{b}(z_t(h)) \frac{dh_t}{dt} = - \left[\left(\left\langle \frac{dz_t(h)}{dt}, \nabla \right\rangle \bar{b}\right)(z_t(h))\right] h_t \tag{3.4}$$

(∇ is the nabla operator). So we see that we can choose h_t to be the projection on the first $2d$ coordinates of the solution to the following $4d$-dimensional initial value problem.

$$\frac{d}{dt}\begin{pmatrix} h_t \\ z_t(h) \end{pmatrix} = \begin{pmatrix} -\bar{b}^{-1}(z_t(h))\left[\left(\left\langle \frac{dz_t(h)}{dt}, \nabla \right\rangle \bar{b}\right)(z_t(h))\right]h_t \\ \bar{b}(z_t(h))h_t \end{pmatrix},$$

$$\begin{pmatrix} h_0 \\ z_0(h) \end{pmatrix} = \begin{pmatrix} \left(\bar{b}\left(\begin{pmatrix} x \\ y \end{pmatrix}\right)\right)^{-1}\left(z^{(1)} - \begin{pmatrix} x \\ y \end{pmatrix}\right) \\ x \\ y \end{pmatrix} \quad (3.5)$$

Existence and uniqueness of a solution to this initial value problem can be obtained from the standard theorems because we ensured that the determinant of $\bar{b}$ is bounded away from zero and hence that the right-hand-side of (3.4) is continuously differentiable.

Step 2: For a general positions of x, $x^{(1)}$, y and $y^{(1)}$ observe that we can divide the action into two parts i.e. timesteps of length 0.5 and choose the streamlines of x and y to be piecewise linear and disjoint.

Step 3: Finally we have to note that by Theorems 1.1 (smoothness) and 1.10 (surjectivity) of [7] we have a submersion in h. □

We may hope that the n-point-motion of a d-dimensional IBF has a density on the set $\mathbb{R}_\times^{nd}$ where no two out of the n points coincide. It should be possible to prove this completely in the same way as Theorem 3.3.1 because $\mathbb{R}_\times^{nd}$ is connected and we can ensure that the streamlines do not intersect each other in arbitrary dimensions. The only remaining task is the need to establish the regularity of the diffusion matrix of the n-point motion (and of course to stipulate sufficient differentiability).

Remark: In this last part we returned to the assumption that b is smooth. There is no reason why it should be impossible to relax this assumptions to require a certain number of derivatives to exist. Nevertheless this would incorporate a lot of technical difficulties (and a lot of checking in the literature) and hence we restrict ourselves to the smooth case here.

Remark: Nothing in this chapter excludes the possibility to apply the results to RIFs. Since we did not state the prerequisites in detail for RIFs we just state that $c > 0$ can be replaced by $c < 0$ without any changes in the assertions or in the proofs.

Chapter 4

A Weak Limit Shape Theorem For Planar Isotropic Brownian Flows

It has been shown by various authors under different assumptions that the diameter of a bounded non-trivial set γ under the action of a stochastic flow grows linearly in time. We show that the asymptotic linear expansion speed if properly defined is deterministic in the case of planar IBFs. This means that we show for a two-dimensional isotropic Brownian flow Φ with a positive Lyapunov exponent that there exists a non-random set $\mathcal{B}$ such that we have for $\epsilon > 0$, arbitrary connected $\gamma \subset\subset \mathbb{R}^2$ consisting of at least two different points and arbitrarily large times T that

$$(1-\epsilon)T\mathcal{B} \subset \bigcup_{0 \leq t \leq T} \bigcup_{x \in \gamma} \Phi_{0,t}(x) \subset (1+\epsilon)T\mathcal{B}.$$

The latter means precisely that for any $t > 0$ there is a $T > t$ such that the inclusions above hold.

4.1 Introduction And Preliminaries

The evolution of the diameter of a bounded set under the action of a stochastic flow has been investigated by various authors with different assumptions and scopes. See [10], [9], [35] and [36] for example. The latter show that the diameter grows linearly in time provided the top Lyapunov exponent is non-negative and also give upper and lower bounds on the expansion speed. Nevertheless these bounds turn out to be far from each other in some examples and there is little hope to match these bounds with the methods from e.g. [9] or [36]. We will follow a different approach which first appeared in [13], wherein a class of periodic stochastic flows on $\mathbb{R}^2$ (or stochastic flows on the torus) is considered. [13] develops a similar limit theorem (even with a stronger assertion) using the fact that their model essentially lives on a compact manifold. Although in the first part we sometimes follow the lines of thought of [13], we will see that to get the assertion we will have to replace the methods relying on the assumption of periodicity (which means perfect dependence of particles which are far from each other) on $\mathbb{R}^2$ by different ones. This is done using the invariance properties with respect to time reversal of IBFs. These properties are not shared by the model of [13] and hence are a novelty in the present subject. The chapter is divided into several sections. First we briefly review the important prerequisites from the literature. Afterwards we give the proper definition of the asymptotic linear expansion speed and state the main result, from which the fact, that the asymptotic expansion speed is constant, turns out to be a corollary. We give the proofs of the main results in the last two sections. The first of these is dedicated to the proof of the lower bound i.e. that the expansion is sufficiently fast. Here we also identify the set $\mathcal{B}$ in terms of a stable norm (which is a concept from [13]). We finally finish the proof in the last section by showing that the expansion is sufficiently slow, for which it will turn out to be sufficient to show that the expansion speed is independent of the initial set. We will work in general dimension d where possible. But since several important features of the proof obviously fail in dimensions larger than two the reader might assume that d is always equal to 2.

4.1.1 Chasing Ball Property, LDP For Discrete Supermartingales

The first of the following lemmas states that the distance of a non-trivial set under the action of the flow tends to approach another moving particle (arbitrary non-anticipating movement) provided that the other particle does not move too fast. Therein we use the following definition.

Definition 4.1.1 (non-trivial set).
A subset of $\mathbb{R}^d$ is called non-trivial if it is bounded, connected and consists of at least two different points.

Note that for IBFs the estimates of the local characteristic and the ellipticity bounds of [53] hold. Therefore we may use the following lemma. For $t \geq 0$ denote by $\mathcal{F}_t := \mathcal{F}_{0,t}$ the sigma-field generated by the flow up to time t.

Lemma 4.1.2 (chasing ball lemma).
Let Φ be an IBF with generator M. Then there are functions $G' : [0,\infty) \times [0,\infty) \times [0,\infty) \to [0,\infty)$ and $G'' : [0,\infty) \times [0,\infty) \to [0,\infty)$, such that there is $r_0 > 0$ depending only on b such that we have the following.

1. *For all $s \in [0,\infty)$ the function $G'(\cdot, s, \cdot)$ is continuous, non-increasing with $\lim_{K \to \infty} \lim_{r \to \infty} G'(K, s, r) = 0$.*

2. *For all $s \in [0,\infty)$ $G''(s, \cdot)$ is continuous and $r \in (0, r_0) \Rightarrow G''(s, r) > 0$.*

3. *Let $s > 0$ and $r < r_0$. Let τ be a finite stopping time for the flow and x, y, z $\mathcal{F}_\tau$-measurable random points in $\mathbb{R}^d$ with $\|x - y\| = r$. Define $r_1 := \|x - z\| \wedge \|y - z\|$, $r_2 := \|\Phi_{\tau, \tau+s}(x) - z\| \wedge \|\Phi_{\tau, \tau+s}(y) - z\|$. Then we have*

$$\mathbb{E}\left[r_2 \vee (r_1 - K) \,|\, \mathcal{F}_\tau\right] \leq r_1 + G'(K, s, r_1) - G''(s, r).$$

Proof: [53, Lemma 2.5]. Observe that K does not appear in the original result in [53] but can be obtained by adding it in the proof of (15) on pages 2055 and 2056 of [53] to obtain instead of (15) the estimate $\mathbb{E}\left[\|x_{\tau+s}\| \vee (x^1 + K)\right] - x^1 \leq g(x^1) + \mathbb{E}\left[N \vee (-K)\right]$ with $\lim_{K \to \infty} \mathbb{E}\left[N \vee (-K)\right] = 0$ and by proceeding as in [53] afterwards. □

The next lemma is similar to a large deviation principle (LDP) for supermartingales and will be used repeatedly during the proof of the lower bound.

Lemma 4.1.3 (Markov martingale bound).
Let $(\xi_j : j \in \mathbb{N})$ be a sequence of real-valued random variables with

1. $\mathbb{E}[\xi_{j+1} | \xi_1, \ldots, \xi_j] \leq 0$,

2. $\forall m \in \mathbb{N} : \exists K_m \in \mathbb{R} : \forall j \in \mathbb{N} : \mathbb{E}[|\xi_j|^m] \leq K_m$.

Then we have

$$\forall \epsilon > 0 : \exists \kappa_m^{(1)} = \kappa_m^{(1)}(\epsilon, (K_n)_{n \in \mathbb{N}}) > 0 : \forall n \in \mathbb{N} : \mathbb{P}\left[\sum_{j=1}^n \xi_j \geq \epsilon n\right] \leq \kappa_m^{(1)} n^{-m}.$$

Proof: [13, Lemma 2]. □

4.1.2 Sub-Gaussian Tails And Sublinear Growth

Lemma 4.1.4. We have for any bounded subset γ of $\mathbb{R}^d$ the following.

1. There is a positive constant C_2, such that we have $\mathbb{P}$-a.s.

$$\limsup_{T \to \infty} \left(\sup_{t \in [0,T]} \sup_{x \in \gamma} \frac{1}{T} \|\Phi_t(x)\| \right) \leq C_2.$$

2. There is a $C_3 > 0$ such that for any $T > 0$ we have

$$\limsup_{n \to \infty} \frac{1}{n^2} \log \mathbb{P}\left[\left(\sup_{t \in [0,T]} \sup_{x \in \gamma} \|\Phi_t(x)\| \geq nT\right)\right] \leq -\frac{T}{2C_3^2}.$$

Proof: [35, Theorem 2.1 and Theorem 2.2]. □

Remark: Some of the statements of this section are far more general in the literature than stated here and there might be some checking of assumptions to get the shapes above. For nothing of this is more than simple computations we do not give any details about it.

4.2 Statement Of The Main Results

Theorem 4.2.1 (limit shape theorem).
Let Φ be a planar IBF (i.e. an IBF with state space $\mathbb{R}^2$) such that the two-point motion has

a strictly positive density apart from the diagonal i.e. on $\mathbb{R}^4_\times := \mathbb{R}^4 \setminus \{z \in \mathbb{R}^4 : z_1 = z_3, z_2 = z_4\}$ (cf. 3.3.1). Assume further that the largest Lyapunov exponent $\frac{1}{2}[\beta_N - \beta_L]$ is strictly positive. For any bounded, connected $\gamma \subset \mathbb{R}^2$ consisting of at least two different points we let $\gamma_t := \Phi_t(\gamma)$ and $\mathcal{W}_t(\gamma) := \bigcup_{0 \leq s \leq t} \gamma_s$. Then there exists a deterministic set $\mathcal{B}$ such that we get for any $\epsilon > 0$ the following.

1. There is $\mathbb{P}$-a.s. $0 < T(\gamma, \epsilon) < \infty$, such that for any $t > T(\gamma, \epsilon)$ we have
$$(1 - \epsilon) t \mathcal{B} \subset \mathcal{W}_t(\gamma).$$

2. There is $\mathbb{P}$-a.s. a sequence $(t_k : k \in \mathbb{N}) \subset \mathbb{R}_+$ with $t_k \nearrow \infty$ that fulfils
$$(1 - \epsilon) t_k \mathcal{B} \subset \mathcal{W}_{t_k}(\gamma) \subset (1 + \epsilon) t_k \mathcal{B}.$$

3. We also have
$$\lim_{T \to \infty} \mathbb{P}\left[(1 - \epsilon) T \mathcal{B} \subset \mathcal{W}_T(\gamma) \subset (1 + \epsilon) T \mathcal{B}\right] = 1.$$

Proof: The proof will be given in Sections 4.3 and 4.4. □

Corollary 4.2.2 (expansion speed).
If we define for γ as above the asymptotic linear expansion speed to be
$$\liminf_{T \to \infty} \frac{1}{T} \operatorname{diam}(\mathcal{W}_T(\gamma))$$
then it is independent of γ and a.s. constant.

Proof: This follows directly from Theorem 4.2.1. □

4.3 The Lower Bound

4.3.1 Hitting Time Of Far Away Balls

Sketch Of Proof

Assume that the original set $\gamma \subset \mathbb{R}^d$ is connected, compact and that it consists of at least two different points (the assumption of compactness is made for simplicity and could be omitted). Denote by $\gamma_t := \Phi_t(\gamma)$ the set γ at time t and by $d_t := \operatorname{diam}(\gamma_t)$ its diameter. Further denote for any $R > 0$ by $\tau^R(\gamma, P) := \inf\{t > 0 : \operatorname{dist}(\gamma_t, P) \leq R, d_t \geq 1\}$ the time it takes for γ to reach an R-neighbourhood of $P \in \mathbb{R}^d$ as a large set. In fact it will turn out that $\liminf_{t \to \infty} d_t \geq 1$ a.s.. We call a subset of $\mathbb{R}^d$ large if it is bounded and has diameter at least 1. Due to the results of [53] and [9] we may assume that γ is large (the following will prove that γ will become large a.s. anyway). Note that we make the standing assumption that the top exponent is strictly positive and that $d \geq 2$.

Theorem 4.3.1 (hitting time theorem).
Let $P \in \mathbb{R}^d$, assume that $\gamma \subset \mathbb{R}^d$ is large and define $\bar{r} := 1 \vee \operatorname{dist}(P, \gamma)$. There is a constant $R > 0$ (neither depending on γ nor on P) such that for any $m \in \mathbb{N}$ there is $\kappa_m^{(2)} > 0$ (neither depending on γ, P nor on $\bar{r}$) such that for $\beta > 1$ we have

$$\mathbb{P}\left[\tau^R(\gamma, P) > \kappa_m^{(2)} \beta \bar{r}\right] \leq \kappa_m^{(2)} \beta^{-m} \bar{r}^{-m}.$$

The proof consists of several steps.

1. Construction of a strictly increasing C^2-function $f : [0, \infty) \to \mathbb{R}$ with $\lim_{r \to \infty} f(r) = \infty$ such that $f(\rho_t^{xy})$ is a submartingale for any $x, y \in \mathbb{R}^d$. The drift of this submartingale has to be bounded away from zero.

2. Estimate of the growth of d_t on average.

3. Estimate of the probability of finding γ_t not being large after a long time i.e. $\mathbb{P}\left[\inf\{s : \forall r > s : d_r > 1\} > t\right]$ is to be bounded from above for large t.

4. Establishing a negative upper bound for the „drift" of $r_t := \operatorname{dist}(\gamma_t, P)$ outside the ball $K_R(P) := \{x \in \mathbb{R}^d : |x - P| < R\}$.

5. Estimation of the tails of $\tau^R(\gamma, P)$.

Construction Of f

The first ingredient needed to construct f is the following lemma.

Lemma 4.3.2 (existence of h).
For any $0 < c_8 < c_9 < \infty$, $\delta > 0$ and $-\infty < c_{10} < 0 < c_{11} < \infty$ there is a decreasing C^2-function $h : [c_8, c_9] \to [h(c_9), 0]$ with

4.3 The Lower Bound

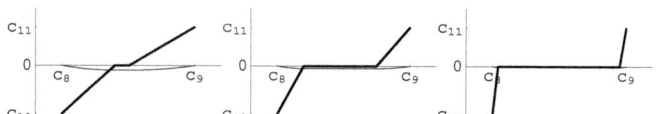

Figure 4.1: h'' (fat) und h' (regular) for some ϵ: 0.99; 0.5; 0.1 (times $0, 5(c_{11} \wedge -c_{10})$)

1. $h'(c_8) = h'(c_9) = 0$,

2. $h''(c_8) = c_{10}$, $h''(c_9) = c_{11}$ and h'' is increasing,

3. $h(c_8) = 0$ and

4. $\sup_{c_8 \leq r \leq c_9}\{|h'(r)|\} \leq \delta$.

Proof: For $0 < \epsilon < 0, 5(c_{11} \wedge -c_{10})$ define $h''_\epsilon : [c_8, c_9] \to [c_{10}, c_{11}]$ via

$$h''_\epsilon(r) := \mathbb{1}_{[c_8, c_{8,\epsilon}]}(r)(r - c_{8,\epsilon})\frac{c_{10}^2}{2\epsilon(c_9 - c_8)} + \mathbb{1}_{[c_{9,\epsilon}, c_9]}(r)(r - c_{9,\epsilon})\frac{c_{11}^2}{2\epsilon(c_9 - c_8)}$$

with $c_{8,\epsilon} := c_8 - \frac{2\epsilon(c_9 - c_8)}{c_{10}}$ and $c_{9,\epsilon} := c_9 - \frac{2\epsilon(c_9 - c_8)}{c_{11}}$ (see Fig. 4.1). Defining

$$h'_\epsilon(r) := \int_{c_8}^r h''_\epsilon(s)ds$$

$$= \mathbb{1}_{[c_8, c_{8,\epsilon}]}(r)\left[(r - c_{8,\epsilon})^2 - (c_8 - c_{8,\epsilon})^2\right]\frac{c_{10}^2}{4\epsilon(c_9 - c_8)} - \mathbb{1}_{[c_{8,\epsilon}, c_9]}(r)\epsilon(c_9 - c_8)$$

$$+ \mathbb{1}_{[c_{9,\epsilon}, c_9]}(r)(r - c_{9,\epsilon})^2\frac{c_{11}^2}{4\epsilon(c_9 - c_8)}$$

ensures 1.. We then also have $h'_\epsilon \leq 0$. 4. follows from choosing $\epsilon \leq \frac{\delta}{c_9 - c_8}$. Setting $h(r) := \int_{c_8}^r h'_\epsilon(s)ds$ for such an ϵ makes h decreasing, ensures 3. and finishes the proof of Lemma 4.3.2. □

Lemma 4.3.3 (existence of f).
There is a strictly increasing C^2-function $f : (0, \infty) \to \mathbb{R}$ with the following properties.

1. $\lim_{r \to \infty} f(r) = \infty$ and $f(\rho_t^{xy})$ is a submartingale for any $x \neq y \in \mathbb{R}^d$.

2. $f(1) = 0$.

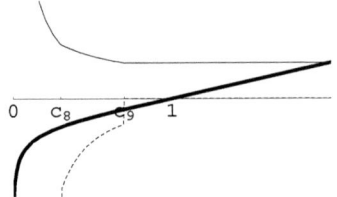

Figure 4.2: f (fat), f' (regular) and f'' (dashed)

3. Writing with Itô's formula, (1.12) and Fubini's theorem

$$\mathbb{E}\left[f(\rho_{t+s}^{xy}) - f(\rho_s^{xy})\right] = \int_s^{t+s} \mathbb{E}\left[f'(\rho_r^{xy})\frac{1-B_N(\rho_r^{xy})}{\rho_r^{xy}}(d-1) + f''(\rho_r^{xy})\left(1-B_L(\rho_r^{xy})\right)\right]dr$$
$$+ \mathbb{E}\left[\int_s^{t+s} f'(\rho_r^{xy})\sqrt{2(1-B_L(\rho_r^{xy}))}\,dW_r\right]$$
$$=: \int_s^{t+s} \mathbb{E}\left[g(\rho_r^{xy})\right]dr + \mathbb{E}\left[\int_s^{t+s} \tilde{g}(\rho_r^{xy})\,dW_r\right]$$

we get that $\tilde{g}$ is bounded and that $g(\cdot) - \frac{1}{8}(\beta_N(d-1) - \beta_L) \geq 0$ on (c_9, ∞).

4. $g : (0, \infty) \to [0, \infty)$ (as above) is continuous and positive.

5. There is $C_8 > 0$ and $C_9 > 0$ such that

$$\mathbb{E}\left[(f(d_{t+1}) - f(d_t)) \wedge C_9 \,|\, \mathcal{F}_t\right] \geq C_8. \tag{4.1}$$

Note that $\mathcal{F}_t$ denotes the σ-field generated by the flow up to time t.

Proof: We choose the following ansatz for f which uses a local linearization of (1.12) near the origin (see Fig. 4.2).

$$f(r) - c_1 := \begin{cases} \log r + c_2 & : 0 < r < c_8 \\ c_3\sqrt{r} + h(r) & : c_8 \leq r \leq c_9 \\ c_4 r + c_5 & : c_9 < r \end{cases}.$$

4.3 The Lower Bound

Put

$$\epsilon := 1 \wedge \frac{1}{8} \frac{\beta_N(d-1) - \beta_L}{d-1} \wedge \frac{1}{24}(\beta_N(d-1) - \beta_L) \left(\frac{\beta_L}{\beta_N(d-1)}\right)^{\frac{1}{3}}$$

$$\wedge \frac{1}{24}(\beta_N(d-1) - \beta_L) \left(\frac{\beta_N(d-1)}{\beta_L}\right)^{-\frac{4}{3}}$$

and choose r_ϵ according to (1.7). Further set

$$c_9 := r_\epsilon \wedge 1, \qquad c_8 := c_9 \left(\frac{\beta_N(d-1)}{\beta_L}\right)^{-\frac{2}{3}},$$

$$\delta := \frac{1}{24}(\beta_N(d-1) - \beta_L) \left\|\frac{1 - B_N(\cdot)}{c_8}\right\|_\infty^{-1} (d-1)^{-1},$$

$$c_{11} := \frac{1}{2c_9\sqrt{c_8 c_9}} \quad \text{and} \quad c_{10} := -\frac{1}{2c_8^2}.$$

Choosing h according to Lemma 4.3.2 and

$$c_3 := \frac{2}{\sqrt{c_8}}, \qquad c_2 := 2 - \log(c_8), \qquad c_4 := \frac{c_3}{2\sqrt{c_9}} + h'(c_9) = \frac{1}{\sqrt{c_8 c_9}},$$

$$c_5 := \sqrt{\frac{c_9}{c_8}} + h(c_9) \quad \text{and} \quad c_1 := -c_4 - c_5$$

ensures the C^2-property of f. We get for the derivatives of f that

$$f'(r) := \begin{cases} \frac{1}{r} & : 0 < r < c_8 \\ \frac{c_3}{2\sqrt{r}} + h'(r) & : c_8 \leq r \leq c_9 \\ c_4 & : c_9 < r \end{cases},$$

$$f''(r) := \begin{cases} -\frac{1}{r^2} & : 0 < r < c_8 \\ -\frac{c_3}{4r\sqrt{r}} + h''(r) & : c_8 \leq r \leq c_9 \\ 0 & : c_9 < r \end{cases}.$$

The submartingale property of $f(d_t)$ will follow if we can show that g is non-negative and that $\tilde{g}$ is bounded. To check this let us give a version of f that only uses c_8, c_9 and δ.

$$f(r) + \sqrt{\frac{c_9}{c_8}} + h(c_9) + \frac{1}{\sqrt{c_8 c_9}} = \begin{cases} \log r + 2 - \log(c_8) & : 0 < r < c_8 \\ \frac{2}{\sqrt{c_8}}\sqrt{r} + h(r) & : c_8 \leq r \leq c_9 \\ \frac{r}{\sqrt{c_8 c_9}} + \sqrt{\frac{c_9}{c_8}} + h(c_9) & : c_9 < r \end{cases},$$

$$f'(r) = \begin{cases} \frac{1}{r} & : 0 < r < c_8 \\ \frac{1}{\sqrt{c_8 r}} + h'(r) & : c_8 \leq r \leq c_9 \\ \frac{1}{\sqrt{c_8 c_9}} & : c_9 < r \end{cases},$$

$$f''(r) = \begin{cases} -\frac{1}{r^2} & : 0 < r < c_8 \\ -\frac{1}{2r\sqrt{c_8 r}} + h''(r) & : c_8 \leq r \leq c_9 \\ 0 & : c_9 < r \end{cases}.$$

The computation for the boundedness of $\tilde{g}$ is rather simple.

$$\begin{aligned} |\tilde{g}(r)| &= \left|f'(r)\sqrt{2(1-B_L(r))}\right| = \left|f'(r)\sqrt{2(1-B_L(r))}\right| \left(\mathbb{1}_{\{r \leq c_8\}} + \mathbb{1}_{\{r > c_8\}}\right) \\ &\leq \left|\frac{1}{r}\sqrt{2(1-B_L(r))}\right| \mathbb{1}_{\{r \leq c_8\}} + \left|f'(r)\sqrt{2(1-B_L(r))}\right| \mathbb{1}_{\{r > c_8\}} \\ &\leq \left|\frac{\sqrt{2}}{r}\sqrt{\frac{1}{2}\beta_L r^2 + \left(1 - B_L(r) - \frac{1}{2}\beta_L r^2\right)}\right| \mathbb{1}_{\{r \leq c_8\}} + |f'(c_8)| \left\|\sqrt{2(1-B_L(.))}\right\|_\infty \\ &\leq \sqrt{\beta_L} + \sqrt{2}\mathbb{1}_{\{r \leq c_8\}} \underbrace{\sqrt{r\left|\frac{1 - B_L(r) - \frac{1}{2}\beta_L r^2}{r^3}\right|}}_{\leq 1} + |f'(c_8)| \left\|\sqrt{2(1-B_L(.))}\right\|_\infty \\ &\leq \sqrt{\beta_L} + \sqrt{2} + |f'(c_8)| \left\|\sqrt{2(1-B_L(.))}\right\|_\infty < \infty. \end{aligned}$$

Now we can turn to the estimate on $g(r)$. For $r \geq c_9$ we obviously have $g(r) > 0$ since

4.3 The Lower Bound

$f'(r) > 0$, $f''(r) = 0$ and $B_N(r) < 1$. For $r \leq c_8$ it is sufficient to consider

$$g(r) = \frac{1}{r^2}\left[(1 - B_N(r))(d-1) - (1 - B_L(r))\right]$$

$$= \frac{1}{2}(\beta_N(d-1) - \beta_L) + \frac{1}{r^2}\left[\left(1 - B_N(r) - \frac{1}{2}\beta_N r^2\right)(d-1) - \left(1 - B_L(r) - \frac{1}{2}\beta_L r^2\right)\right]$$

$$\geq \frac{1}{2}(\beta_N(d-1) - \beta_L) - 2\underbrace{r}_{\leq 1}\underbrace{\left[\left|\frac{1 - B_N(r) - \frac{1}{2}\beta_N r^2}{r^3}\right|(d-1) \vee \left|\frac{1 - B_L(r) - \frac{1}{2}\beta_L r^2}{r^3}\right|\right]}_{\leq \frac{1}{8}(\beta_N(d-1) - \beta_L)}$$

$$\geq \frac{1}{4}(\beta_N(d-1) - \beta_L) > 0.$$

The remaining case $c_8 \leq r \leq c_9$ needs a little more attention.

$$g(r) = \left[\frac{1}{\sqrt{c_8 r}} + h'(r)\right]\frac{1 - B_N(r)}{r}(d-1) + \left[-\frac{1}{2r\sqrt{c_8 r}} + h''(r)\right](1 - B_L(r))$$

$$= \left[\frac{1}{\sqrt{c_8 r}} + h'(r)\right]\left(\frac{1}{2}\beta_N r(d-1) + \frac{1 - B_N(r) - \frac{1}{2}\beta_N r^2}{r}(d-1)\right)$$

$$+ \left[h''(r) - \frac{1}{2r\sqrt{c_8 r}}\right]\left[\frac{1}{2}\beta_L r^2 + \left(1 - B_L(r) - \frac{1}{2}\beta_L r^2\right)\right]$$

$$= \frac{\beta_N r}{2\sqrt{c_8 r}}(d-1) - \frac{\beta_L r^2}{2}\left(\frac{1}{2r\sqrt{c_8 r}} - h''(r)\right) + \frac{1}{\sqrt{c_8 r}}\left(\frac{1 - B_N(r) - \frac{1}{2}\beta_N r^2}{r}(d-1)\right)$$

$$+ \left(h''(r) - \frac{1}{2r\sqrt{c_8 r}}\right)\left(1 - B_L(r) - \frac{1}{2}\beta_L r^2\right) + h'(r)\frac{1 - B_N(r)}{r}(d-1)$$

$$\geq \underbrace{\frac{\beta_N r}{2\sqrt{c_8 r}}(d-1) - \frac{\beta_L r^2}{2}\left(\frac{1}{2r\sqrt{c_8 r}} + \frac{1}{2c_8^2}\right)}_{=:I} - \underbrace{\frac{1}{\sqrt{c_8 r}}\left|\frac{1 - B_N(r) - \frac{1}{2}\beta_N r^2}{r}(d-1)\right|}_{=:II}$$

$$- \underbrace{\left|h''(r) - \frac{1}{2r\sqrt{c_8 r}}\right|\left|1 - B_L(r) - \frac{1}{2}\beta_L r^2\right|}_{=:III} - \underbrace{\left|h'(r)\frac{1 - B_N(r)}{r}(d-1)\right|}_{=:IV}.$$

For we chose $c_8 := c_9 \left(\frac{\beta_N(d-1)}{\beta_L}\right)^{-\frac{2}{3}}$ and $c_9 := r_\epsilon \wedge 1$ we get

$$I = \frac{\beta_N r}{2\sqrt{c_8 r}}(d-1) - \frac{\beta_L r^2}{2}\left(\frac{1}{2r\sqrt{c_8 r}} + \frac{1}{2c_8^2}\right)$$

$$= \frac{\sqrt{r}}{2\sqrt{c_8}}\left[\frac{1}{2}(\beta_N(d-1) - \beta_L) + \frac{1}{2}\left(\beta_N(d-1) - r^{\frac{3}{2}} c_8^{-\frac{3}{2}} \beta_L\right)\right]$$

$$\geq \frac{1}{4}(\beta_N(d-1) - \beta_L) + \frac{1}{4}\sqrt{\frac{r}{c_8}}\left[\beta_N(d-1) - c_9^{\frac{3}{2}} c_8^{-\frac{3}{2}} \beta_L\right]$$

$$= \frac{1}{4}(\beta_N(d-1) - \beta_L),$$

$$II \leq \frac{r\sqrt{rc_9}}{\sqrt{c_8 r}}\left|\frac{1 - B_N(r) - \frac{1}{2}\beta_N r^2}{r^3}(d-1)\right| \leq \sqrt{\frac{c_9}{c_8}}\left|\frac{1 - B_N(r) - \frac{1}{2}\beta_N r^2}{r^3}(d-1)\right|$$

$$\leq \frac{1}{24}(\beta_N(d-1) - \beta_L),$$

$$III = \left|h''(r) - \frac{1}{2r\sqrt{c_8 r}}\right|\left|1 - B_L(r) - \frac{1}{2}\beta_L r^2\right| \leq \left|\frac{1}{2c_8^2} + \frac{1}{2r\sqrt{c_8 r}}\right| r^3 \left|\frac{1 - B_L(r) - \frac{1}{2}\beta_L r^2}{r^3}\right|$$

$$\leq \frac{c_9^2}{c_8^2} r \left|\frac{1 - B_L(r) - \frac{1}{2}\beta_L r^2}{r^3}\right| \leq \left(\frac{\beta_N(d-1)}{\beta_L}\right)^{\frac{4}{3}} \left|\frac{1 - B_L(r) - \frac{1}{2}\beta_L r^2}{r^3}\right|$$

$$\leq \frac{1}{24}(\beta_N(d-1) - \beta_L).$$

Finally since $\delta = \frac{1}{24}(\beta_N(d-1) - \beta_L) \left\|\frac{1-B_N(.)}{c_8}\right\|_\infty^{-1} (d-1)^{-1}$ we get

$$IV = \left|h'(r)\frac{1 - B_N(r)}{r}(d-1)\right| \leq \frac{1}{24}(\beta_N(d-1) - \beta_L)$$

and

$$g(r) \geq (\beta_N(d-1) - \beta_L)\left[\frac{1}{4} - \frac{1}{24} - \frac{1}{24} - \frac{1}{24}\right] = \frac{1}{8}(\beta_N(d-1) - \beta_L) > 0.$$

It remains to show (4.1). This is done in the following subsubsection.

Growth Of $f(d_t)$ On Average

There are two cases. If $d_t < \frac{r^{(\epsilon)}}{2}$ it is sufficient to consider the two-point motion. Due to the Markov-property of the submartingale $f(\rho_t^{xy})$ we get choosing (adaptedly) x_t and y_t

4.3 THE LOWER BOUND

Figure 4.3: growth of d_t on average

with $|x_t - y_t| = d_t$ and some constant $C_9 > 0$ (to be specified later)

$$\mathbb{E}\left[((f(d_{t+1}) - f(d_t)) \wedge C_9) \mathbb{1}_{\{d_t < \frac{r^{(\epsilon)}}{2}\}} \Big| \mathcal{F}_t\right]$$
$$\geq \mathbb{E}\left[(f(\rho_{t+1}^{xy}) - f(\rho_t^{xy})) \wedge C_9 | \mathcal{F}_t\right] \mathbb{1}_{\{f(d_t) < f\left(\frac{r^{(\epsilon)}}{2}\right)\}}$$
$$= \mathbb{E}_{f(\rho_t^{xy})}\left[(f(\rho_1^{xy}) - f(\rho_0^{xy})) \wedge C_9\right] \mathbb{1}_{\{f(d_t) < f\left(\frac{r^{(\epsilon)}}{2}\right)\}}$$
$$\geq \left[\left(f\left(r^{(\epsilon)}\right) - f\left(\frac{r^{(\epsilon)}}{2}\right)\right) \wedge C_9\right] \mathbb{P}_{f(\rho_t^{xy})}\left[\sup_{0 \leq s \leq 1} f(\rho_s^{xy}) \geq f(r^{(\epsilon)})\right] \mathbb{1}_{\{f(d_t) < f\left(\frac{r^{(\epsilon)}}{2}\right)\}}$$
$$+ \left(\frac{1}{8}(\beta_N(d-1) - \beta_L) \wedge C_9\right) \mathbb{P}_{f(\rho_t^{xy})}\left[\sup_{0 \leq s \leq 1} f(\rho_s^{xy}) < f(r^{(\epsilon)})\right] \mathbb{1}_{\{f(d_t) < f\left(\frac{r^{(\epsilon)}}{2}\right)\}}$$
$$\geq \left(\left[f\left(r^{(\epsilon)}\right) - f\left(\frac{r^{(\epsilon)}}{2}\right)\right] \wedge \frac{1}{8}(\beta_N(d-1) - \beta_L) \wedge C_9\right) \mathbb{1}_{\{d_t < \frac{r^{(\epsilon)}}{2}\}}.$$

In case $d_t \geq \frac{r^{(\epsilon)}}{2}$ first consider the growth of d_t. We may assume $G''(1, \frac{r^{(\epsilon)}}{10}) =: \frac{c_6}{c_4} > 0$ (otherwise we decrease $r^{(\epsilon)}$, see Lemma 4.1.2). There are $\hat{r}$ and $C_9 > 0$ such that for any $r \geq \hat{r}$ we have $G'(\frac{C_9}{2c_4}, 1, r) < \frac{c_6}{2c_4}$. Choose (adaptedly) $x_t^{(1)}, x_t^{(2)}, y_t^{(1)}, y_t^{(2)} \in \gamma_t$ with $|x_t^{(1)} - x_t^{(2)}| = d_t, |x_t^{(i)} - y_t^{(i)}| = \frac{r^{(\epsilon)}}{10} : i = 1, 2$ and define

$$z^{(1)} := x_t^{(1)} + \frac{x_t^{(1)} - x_t^{(2)}}{|x_t^{(1)} - x_t^{(2)}|}\hat{r}, z^{(2)} := x_t^{(2)} + \frac{x_t^{(2)} - x_t^{(1)}}{|x_t^{(2)} - x_t^{(1)}|}\hat{r},$$
$$r_1^{(i)} := |x_t^{(i)} - z^{(i)}| \wedge |y_t^{(i)} - z^{(i)}| = \hat{r} : i = 1, 2 \text{ and}$$
$$r_2^{(i)} := |x_{t+1}^{(i)} - z^{(i)}| \wedge |y_{t+1}^{(i)} - z^{(i)}| : i = 1, 2$$

(see Fig. 4.3 for the geometry at time t). Lemma 4.1.2 provides for $i = 1, 2$ that we have

$$\mathbb{E}\left[(c_4(r_1^{(i)} - r_2^{(i)})) \wedge \tfrac{C_9}{2}\Big| \mathcal{F}_t\right] \geq G''(1, \tfrac{r^{(e)}}{10}) - G'\left(\tfrac{C_9}{2c_4}, 1, \hat{r}\right) \geq \tfrac{c_6}{2} > 0 \text{ and therefore}$$

$$|z^{(1)} - z^{(2)}| = r_1^{(1)} + d_t + r_1^{(2)} \text{ and } |z^{(1)} - z^{(2)}| \leq r_2^{(1)} + d_{t+1} + r_2^{(2)} \Rightarrow$$

$$(c_4(d_{t+1} - d_t)) \wedge C_9 \geq (c_4(r_1^{(1)} - r_2^{(1)})) \wedge \frac{C_9}{2} + (c_4(r_1^{(2)} - r_2^{(2)})) \wedge \frac{C_9}{2} \Rightarrow$$

$$\mathbb{E}[(c_4(d_{t+1} - d_t)) \wedge C_9 |\mathcal{F}_t] \geq \mathbb{E}\left[(c_4(r_1^{(1)} - r_2^{(1)})) \wedge \tfrac{C_9}{2}\Big| \mathcal{F}_t\right] + \mathbb{E}\left[(c_4(r_1^{(2)} - r_2^{(2)})) \wedge \tfrac{C_9}{2}\Big| \mathcal{F}_t\right] \geq c_6.$$

Now we turn this into an estimate for $f(d_t)$. Abbreviate $\rho_t := \rho_t^{x^{(1)}x^{(2)}}$ and consider for $K > 0$

$$\mathbb{E}\left[((f(d_{t+1}) - f(d_t)) \wedge C_9) \mathbb{1}_{\left\{f\left(\tfrac{r^{(e)}}{2}\right) \leq f(d_t) \leq K\right\}}\Big| \mathcal{F}_t\right]$$

$$\geq \mathbb{E}\left[((f(\rho_{t+1}) - f(\rho_t)) \wedge C_9) \mathbb{1}_{\left\{f\left(\tfrac{r^{(e)}}{2}\right) \leq f(d_t) \leq K\right\}}\Big| \mathcal{F}_t\right] \quad (4.2)$$

$$\geq \inf_{f\left(\tfrac{r^{(e)}}{2}\right) \leq r \leq K} \mathbb{E}_{f(\rho_0) = r}\left[(f(\rho_1) - f(\rho_0)) \wedge C_9\right] \mathbb{1}_{\left\{f\left(\tfrac{r^{(e)}}{2}\right) \leq f(d_t) \leq K\right\}} =: \mathbb{1}_{\left\{f\left(\tfrac{r^{(e)}}{2}\right) \leq f(d_t) \leq K\right\}} c_7 > 0.$$

The last inequality follows from the continuity and positivity ($g(r) > 0$ for $r \geq 0$) of the mapping $r \mapsto \mathbb{E}_{f(\rho_0)=r}\left[(f(\rho_1) - f(\rho_0)) \wedge C_9\right]$. (4.1) is now an easy consequence of the following proposition.

Proposition 4.3.4 (lower drift bound).
There is $K \in \mathbb{N}$ such that $\mathbb{E}\left[((f(d_{t+1}) - f(d_t)) \wedge C_9) \mathbb{1}_{\{f(d_t) > K\}}\big|\mathcal{F}_t\right] \geq \tfrac{c_6}{2} \mathbb{1}_{\{f(d_t) > K\}}$.

Proof of Proposition 4.3.4: Consider for $F \in \mathcal{F}_t$ and $K \in \mathbb{N}$ that

$$\mathbb{E}\left[((f(d_{t+1}) - f(d_t)) \wedge C_9) \mathbb{1}_{\{f(d_t) > K\}} \mathbb{1}_F\right]$$
$$= \mathbb{E}\left[((f(d_{t+1}) - f(d_t)) \wedge C_9) \mathbb{1}_{\{f(d_t) > K\} \cap F \cap \{f(d_{t+1}) \geq 0\}}\right]$$
$$+ \mathbb{E}\left[((f(d_{t+1}) - f(d_t)) \wedge C_9) \mathbb{1}_{\{f(d_t) > K\} \cap F \cap \{f(d_{t+1}) < 0\}}\right]$$
$$\geq \mathbb{E}\left[((c_4(d_{t+1} - d_t)) \wedge C_9) \mathbb{1}_{\{f(d_t) > K\} \cap F \cap \{f(d_{t+1}) \geq 0\}}\right]$$
$$+ \mathbb{E}\left[(f(\rho_{t+1}) - f(\rho_t)) \mathbb{1}_{\{f(\rho_{t+1}) - f(\rho_t) < -K\}} \mathbb{1}_{\{f(d_t) > K\} \cap F \cap \{f(d_{t+1}) < 0\}}\right]$$
$$\geq \mathbb{E}\left[((c_4(d_{t+1} - d_t)) \wedge C_9) \mathbb{1}_{\{f(d_t) > K\} \cap F}\right]$$
$$+ \mathbb{E}\left[\mathbb{E}_{f(\rho_t)}\left[(f(\rho_1) - f(\rho_0)) \mathbb{1}_{\{f(\rho_1) - f(\rho_0) < -K\}}\right] \mathbb{1}_{\{f(d_t) > K\} \cap F}\right]$$

4.3 THE LOWER BOUND

$$\geq \mathbb{E}\left[\mathbb{E}\left[(c_4(d_{t+1}-d_t))\wedge C_9\,|\,\mathcal{F}_t\right]\mathbb{1}_{\{f(d_t)>K\}\cap F}\right]$$
$$-\mathbb{E}\left[\mathbb{1}_{\{f(d_t)>K\}\cap F}\sum_{n=K}^{\infty}n\mathbb{P}_{f(\rho_t)}\left[f(\rho_1)-f(\rho_0)<1-n\right]\right] \qquad (4.3)$$
$$=:\mathbb{E}\left[\mathbb{E}\left[(c_4(d_{t+1}-d_t))\wedge C_9\,|\,\mathcal{F}_t\right]\mathbb{1}_{\{f(d_t)>K\}\cap F}\right]-I \geq c_6\mathbb{P}\left[f(d_t)>K;F;f(d_{t+1})\geq 0\right]-I.$$

To get an estimatie on I the next lemma is useful.

Lemma 4.3.5 (tails for shrinking).
For $x \neq y \in \mathbb{R}^2$ and $r \in \mathbb{R}$ we have $\mathbb{P}_{f(\rho_t^{xy})=r}\left[f(\rho_0^{xy})-f(\rho_1^{xy})>n\right] \leq \frac{\|\tilde{g}\|_\infty}{n\sqrt{2\pi}}\exp\left\{-\frac{n^2}{2\|\tilde{g}\|_\infty^2}\right\}.$

Proof of Lemma 4.3.5: Due to some standard results ([21, Proposition 5.2.18], [21, Theorem 4.6] and [46, Chapter V, Theorem 1.7] e.g.) and due to $df(\rho_s^{xy}) = \underbrace{\tilde{g}(\rho_s^{xy})}_{\leq \|\tilde{g}\|_\infty}dW_s + \underbrace{g(\rho_s^{xy})}_{\geq 0}ds$
we see that

$$\mathbb{P}_{f(\rho_t^{xy})=r}\left[f(\rho_0^{xy})-f(\rho_1^{xy})>n\right] \leq \mathbb{P}_{f(\rho_t^{xy})=r}\left[\int_0^1 \tilde{g}(\rho_s^{xy})dW_s < -n\right]$$
$$= \mathbb{P}_{f(\rho_t^{xy})=r}\left[W_{\langle\int_0^\cdot \tilde{g}(\rho_s^{xy})dW_s\rangle_1} < -n\right] \leq \mathbb{P}\left[W_{\int_0^1 \|\tilde{g}\|_\infty^2 ds} < -n\right] \leq \mathbb{P}\left[\|\tilde{g}\|_\infty W_1 < -n\right]$$
$$= 1 - F_{\mathcal{N}(0,1)}\left(\frac{n}{\|\tilde{g}\|_\infty}\right) \leq \frac{\|\tilde{g}\|_\infty}{n\sqrt{2\pi}}\exp\left\{-\frac{n^2}{2\|\tilde{g}\|_\infty^2}\right\}. \qquad \square$$

The conclusion $I \leq \mathbb{E}\left[\mathbb{1}_{\{f(d_t)>K\}\cap F}\sum_{n=K}^{\infty}\frac{n\|\tilde{g}\|_\infty}{(n-1)\sqrt{2\pi}}\exp\left\{-\frac{(n-1)^2}{2\|\tilde{g}\|_\infty^2}\right\}\right] \stackrel{K\to\infty}{\to} 0$ implies for sufficiently large K (uniformly in F) that $I \leq \frac{c_4 c_6}{2}\mathbb{P}\left[\{f(d_t)>K\}\cap F\cap\{f(d_{t+1})\geq 0\}\right]$, because $\mathbb{P}\left[f(d_{t+1})\geq 0\,|\,f(d_t)>K\right] \to 1$ for $K\to\infty$ which together with (4.3) completes the proof of Proposition 4.3.4. $\square$

The proof of (4.1) is now straightforward. Choose $K>1$ for which Proposition 4.3.4 holds, $c_7 = c_7(K)$ according to (4.2) and consider

$$\mathbb{E}\left[(f(d_{t+1})-f(d_t))\wedge C_9\,|\,\mathcal{F}_t\right]$$
$$=\mathbb{E}\left[((f(d_{t+1})-f(d_t))\wedge C_9)\,\mathbb{1}_{\left\{f(d_t)<f\left(\frac{r(e)}{2}\right)\right\}}\,\Big|\,\mathcal{F}_t\right]$$
$$+\mathbb{E}\left[((f(d_{t+1})-f(d_t))\wedge C_9)\,\mathbb{1}_{\left\{f\left(\frac{r(e)}{2}\right)\leq f(d_t)\leq K\right\}}\,\Big|\,\mathcal{F}_t\right]$$
$$+\mathbb{E}\left[((f(d_{t+1})-f(d_t))\wedge C_9)\,\mathbb{1}_{\{f(d_t)>K\}}\,\Big|\,\mathcal{F}_t\right]$$

$$\geq \left(\left[f(r^{(\epsilon)}) - f(\tfrac{r_\epsilon}{2}) \right] \wedge \tfrac{1}{8}(\beta_N(d-1) - \beta_L) \wedge C_9 \right) \mathbb{1}_{\left\{ f(d_t) < \frac{r^{(\epsilon)}}{2} \right\}}$$
$$+ c_7 \mathbb{1}_{\left\{ \frac{r^{(\epsilon)}}{2} \leq f(d_t) \leq K \right\}} + \tfrac{c_6}{2} \mathbb{1}_{\{f(d_t) > K\}}$$
$$\geq \left[f\left(r^{(\epsilon)}\right) - f\left(\tfrac{r^{(\epsilon)}}{2}\right) \right] \wedge \tfrac{1}{8}(\beta_N(d-1) - \beta_L) \wedge C_9 \wedge c_7 \wedge \tfrac{c_6}{2} =: C_8 > 0.$$

The proof of Lemma 4.3.3 is complete. □

The estimate (on average) is about to be transformed into one of the probability of the event that our original set is not large after a long time.

Pathwise Growth Of $f(d_t)$

We have $\mathbb{E}\left[f(d_{t+1}) \wedge (f(d_t) + C_9) \,|\, \mathcal{F}_t\right] - f(d_t) \geq C_8 > 0$. So we can verify the assumptions of Lemma 4.1.3 for $\xi_i := f(d_{i-1}) - [f(d_i) \wedge (f(d_{i-1}) + C_9)] + C_8$. We only have to prove $\mathbb{E}[|\xi_i|^m] \leq K_m$ for certain real K_m.

$$\begin{aligned}
\mathbb{E}\left[|\xi_i|^m\right] &= \mathbb{E}\left[|f(d_{i-1}) - [f(d_i) \wedge (f(d_{i-1}) + C_9)] + C_8|^m\right] \\
&\leq 2^m C_8^m + 2^m \mathbb{E}\left[|f(d_{i-1}) - [f(d_i) \wedge (f(d_{i-1}) + C_9)]|^m\right] \\
&\leq 2^m C_8^m + 2^m \mathbb{E}\left[|f(d_{i-1}) - [f(d_i) \wedge (f(d_{i-1}) + C_9)]|^m \mathbb{1}_{\{d_i < d_{i-1}\}}\right] \\
&\quad + 2^m \mathbb{E}\left[|f(d_{i-1}) - [f(d_i) \wedge (f(d_{i-1}) + C_9)]|^m \mathbb{1}_{\{d_{i-1} \leq d_i\}}\right] \\
&\leq 2^m C_8^m + 2^m C_9^m \mathbb{P}\left[d_i \geq d_{i-1}\right] + 2^m \mathbb{E}\left[(f(d_{i-1}) - f(d_i))^m \mathbb{1}_{\{d_{i-1} > d_i\}}\right] \\
&\leq 2^m C_8^m + 2^m C_9^m + 2^m \underbrace{\mathbb{E}\left[(f(d_{i-1}) - f(d_i))^m \mathbb{1}_{\{d_{i-1} > d_i\}}\right]}_{=: I}
\end{aligned} \quad (4.4)$$

To estimate I we choose x and y (in an adapted way!) in γ such that $\|x_{i-1} - y_{i-1}\| = d_{i-1}$. For shrinking of d_t implies decreasing of the distance of x_t and y_t we can further conclude

$$\begin{aligned}
I &= \mathbb{E}\left[(f(d_{i-1}) - f(d_i))^m \mathbb{1}_{\{d_{i-1} > d_i\}}\right] \leq \mathbb{E}\left[(f(\rho_{i-1}^{xy}) - f(\rho_i^{xy}))^m \mathbb{1}_{\{\rho_{i-1}^{xy} > \rho_i^{xy}\}}\right] \\
&= \mathbb{E}\left[\mathbb{E}\left[(f(\rho_{i-1}^{xy}) - f(\rho_i^{xy}))^m \mathbb{1}_{\{f(\rho_{i-1}^{xy}) > f(\rho_i^{xy})\}} \,\Big|\, \mathcal{F}_{i-1}\right]\right] \\
&\leq \mathbb{E}\left[\mathbb{E}_{f(\rho_{i-1}^{xy})}\left[(f(\rho_0^{xy}) - f(\rho_1^{xy}))^m \mathbb{1}_{\{f(\rho_0^{xy}) > f(\rho_1^{xy})\}}\right]\right] \\
&\leq 1 + \sum_{n=1}^{\infty} (n+1)^m \sup_{f(r) \in \mathbb{R}^+} \mathbb{P}_{f(r)}\left[f(\rho_0^{xy}) - f(\rho_1^{xy}) > n\right].
\end{aligned} \quad (4.5)$$

4.3 THE LOWER BOUND

Combining (4.4), (4.5) and Lemma 4.3.5 yields

$$\mathbb{E}\left[|\xi_i|^m\right] \leq 2^m(C_8^m + C_9^m + 1) + 2^m \sum_{n=1}^{\infty} \frac{\|\tilde{g}\|_\infty (n+1)^m}{n\sqrt{2\pi}} \exp\left\{-\frac{n^2}{2\|\tilde{g}\|_\infty^2}\right\} =: K_m.$$

Concluding with Lemma 4.1.3 we have for $m \in \mathbb{N}$ the existence of $\kappa_m^{(1)} \in \mathbb{R}$, such that for $n \geq \frac{2}{C_8}|f(d_0)|$ the following holds.

$$\begin{aligned}
\mathbb{P}\left[d_n < 1\right] &= \mathbb{P}\left[f(d_n) < 0\right] = \mathbb{P}\left[\sum_{i=0}^{n-1} \left(f(d_i) - f(d_{i+1}) + C_8\right) > f(d_0) + C_8 n\right] \\
&\leq \mathbb{P}\left[\sum_{i=0}^{n-1} \left(f(d_i) - [f(d_{i+1}) \wedge (f(d_i) + C_9)] + C_8\right) \geq f(d_0) + C_8 n\right] \\
&= \mathbb{P}\left[\sum_{i=1}^{n} \xi_i \geq f(d_0) + C_8 n\right] \leq \mathbb{P}\left[\sum_{i=1}^{n} \xi_i \geq \frac{C_8 n}{2}\right] \leq \kappa_m^{(1)} n^{-m}. \quad (4.6)
\end{aligned}$$

Increasing $\kappa_m^{(1)}$ if necessary ensures that this holds for all n (this is to be assumed).
Remark: The assumption of largeness of γ makes this correction uniform in γ.
So we can estimate the probability of $F_n := \{\exists i \in [\lfloor\sqrt{n}\rfloor, \infty] \cap \mathbb{N} : d_i < 1\}$ via

$$\mathbb{P}\left[F_n\right] \leq \sum_{i=\lfloor\sqrt{n}\rfloor}^{\infty} \mathbb{P}\left[d_i < 1\right] \leq \sum_{i=\lfloor\sqrt{n}\rfloor}^{\infty} \kappa_{2+2m}^{(1)} i^{-2-2m} \leq \underbrace{\left(\kappa_{2+2m}^{(1)} \sum_{i=1}^{\infty} i^{-2}\right)}_{=:\kappa_m^{(6)}} n^{-m}. \quad (4.7)$$

A simple Borel-Cantelli argument shows that the flow cannot contract a non-trivial set to a point i.e. d_t a.s. does not converge to zero as $t \to \infty$.

Estimates On The Tails Of $\tau^R(\gamma, P)$

First let $r_t := \text{dist}(\gamma_t, P)$ and observe for $n \in \mathbb{N}$ that we have the estimate

$$\mathbb{P}\left[\tau^R(\gamma, P) > n\right] \leq \mathbb{P}\left[\bigcap_{i=\lfloor\sqrt{n}\rfloor}^{n} (\{r_i > R\} \cup \{d_i < 1\})\right] \leq \mathbb{P}\left[F_n\right] + \mathbb{P}\left[F_n^C, \bigcap_{i=\lfloor\sqrt{n}\rfloor}^{n} \{r_i > R\}\right]$$

$$=: I + II. \quad (4.8)$$

I is already treated, so only II is left. For arbitrary $\delta > 0$ and $n \geq 4 \vee 4(r_0 - R)\delta^{-1}$ we can estimate

$$II \leq \mathbb{P}\left[\bigcap_{i=\lfloor n \rfloor}^{n} \{r_i > R\}, \bigcap_{i=\lfloor \sqrt{n} \rfloor}^{\infty} \{d_i \geq 1\}\right]$$

$$= \mathbb{P}\left[(r_{\lfloor \sqrt{n} \rfloor} - r_0) + \underbrace{\sum_{i=\lfloor \sqrt{n} \rfloor+1}^{n} (r_i - r_{i-1})}_{\geq -\frac{\delta}{2}(n-\sqrt{n})} > R - r_0, \bigcap_{i=\lfloor n \rfloor}^{n-1} \{r_i > R\}, \bigcap_{i=\lfloor \sqrt{n} \rfloor}^{\infty} \{d_i \geq 1\}\right]$$

$$\leq \mathbb{P}\left[\eta^{(n)} + \sum_{i=\lfloor \sqrt{n} \rfloor+1}^{n} \left[(r_i - r_{i-1})\mathbb{1}_{\{d_{i-1} \geq 1, r_{i-1} > R\}} - \delta\mathbb{1}_{\{d_{i-1} < 1\} \cup \{r_{i-1} \leq R\}} + \delta\right] > \underbrace{\frac{\delta}{2}(n - \lfloor \sqrt{n} \rfloor)}_{\geq \frac{\delta n}{4}}\right]$$

$$\leq \mathbb{P}\left[\eta^{(n)} \geq \frac{\delta n}{8}\right]$$

$$+ \mathbb{P}\left[\sum_{i=\lfloor \sqrt{n} \rfloor+1}^{n} \left[(r_i - r_{i-1})\mathbb{1}_{\{d_{i-1} \geq 1, r_{i-1} > R\}} - \delta\mathbb{1}_{\{d_{i-1} < 1\} \cup \{r_{i-1} \leq R\}} + \delta\right] \geq \frac{\delta}{4}(n - \lfloor \sqrt{n} \rfloor)\right]$$

$$=: III + IV. \tag{4.9}$$

Therein $\eta^{(n)} := r_{\lfloor \sqrt{n} \rfloor} - r_0$ is used. The term III can be estimated by the growth of a Brownian motion. Choose $z \in \gamma$ with $\|z - P\| = r_0$. Then we have

$$III \leq \mathbb{P}\left[\left\|z_{\lfloor \sqrt{n} \rfloor} - z_0\right\| \geq \frac{\delta n}{8}\right] \leq \mathbb{P}\left[\exists 1 \leq i \leq d : \left|z^i_{\lfloor \sqrt{n} \rfloor} - z^i_0\right| \geq \frac{\delta n}{8d}\right]$$

$$\leq d\mathbb{P}\left[n^{\frac{1}{4}}\left|z^1_1 - z^1_0\right| \geq \frac{\delta n}{8d}\right] = d \int_{\frac{\delta}{8d}n^{\frac{3}{4}}}^{\infty} \frac{2}{\sqrt{2\pi}} e^{-\frac{t^2}{2}} dt \leq \kappa_m^{(3)} n^{-m} \tag{4.10}$$

for suitable $\kappa_m^{(3)} \in \mathbb{R}$. The estimation of IV applies Lemma 4.1.3 again. For $\delta > 0$ and $n \geq 4 \vee 4(r_0 - R)\delta^{-1}$ observe

$$IV \leq \mathbb{P}\left[\sum_{i=\lfloor \sqrt{n} \rfloor+1}^{n} \xi_i^{(C_{10},\delta)} \geq (n - \lfloor \sqrt{n} \rfloor)\frac{\delta}{4}\right]. \tag{4.11}$$

4.3 THE LOWER BOUND

Therein for $C_{10} > 0$ and $i \in \mathbb{N}$ set

$$\xi_i^{(C_{10},\delta)} := \left[(r_i \vee (r_{i-1} - C_{10}) - r_{i-1}) \mathbb{1}_{\{d_{i-1} \geq 1, r_{i-1} > R\}} - \delta \mathbb{1}_{\{d_{i-1} < 1\} \cup \{r_{i-1} \leq R\}} + \delta \right].$$

The sequel aims at showing that $(\xi_i^{(C_{10},\delta)} : i \in \mathbb{N})$ for suitable C_{10} and δ satisfies the assumptions of Lemma 4.1.3. Afterwards this lemma and a treatment of the fact that there are some terms $\xi_i^{(C_{10},\delta)}$ missing in the last sum which makes Lemma 4.1.3 not directly suitable for (4.11) will complete the proof. Therefore we have to show $\mathbb{E}\left[\left|\xi_i^{(C_{10},\delta)}\right|^m\right] \leq K_m < \infty$ for any m and uniformly in i.

$$\begin{aligned}
\mathbb{E}\left[\left|\xi_i^{(C_{10},\delta)}\right|^m\right] &\leq 2^m \delta^m + 2^m \mathbb{E}\left[|r_i \vee (r_{i-1} - C_{10}) - r_{i-1}|^m \mathbb{1}_{\{d_{i-1} \geq 1\}}\right] \\
&= 2^m \delta^m + 2^m \mathbb{E}\left[|r_i \vee (r_{i-1} - C_{10}) - r_{i-1}|^m \mathbb{1}_{\{d_{i-1} \geq 1, r_i < r_{i-1}\}}\right] \\
&\quad + 2^m \mathbb{E}\left[|r_i \vee (r_{i-1} - C_{10}) - r_{i-1}|^m \mathbb{1}_{\{d_{i-1} \geq 1, r_i > r_{i-1}\}}\right] \\
&\leq 2^m \delta^m + 2^m C_{10}^m + 2^m \underbrace{\mathbb{E}\left[(r_i - r_{i-1})^m \mathbb{1}_{\{d_{i-1} \geq 1, r_i > r_{i-1}\}}\right]}_{=:V}.
\end{aligned}$$

For γ cannot get away from P without having its nearest (w.r.t. P) point doing so we can proceed for the estimation of V as follows. Let $z \in \gamma$ such that $\|z_{i-1} - P\| = r_{i-1}$ and consider

$$V = \mathbb{E}\left[(r_i - r_{i-1})^m \mathbb{1}_{\{d_{i-1} \geq 1, r_i > r_{i-1}\}}\right] \leq \mathbb{E}\left[|z_i - z_{i-1}|^m\right] = \mathbb{E}\left[\left|\mathcal{N}(0,1)^{\otimes d}\right|^m\right].$$

Therefore we can choose $K_m = K_m(C_{10}, \delta) := 2^m \delta^m + 2C_{10}^m + 2^m \mathbb{E}\left[\left|\mathcal{N}(0,1)^{\otimes d}\right|^m\right]$ and it only remains to show that there are $C_{10} > 0$ and $\delta > 0$ such that $\mathbb{E}\left[\xi_i^{(C_{10},\delta)} \middle| \xi_{i-1}^{(C_{10},\delta)} \cdots \xi_1^{(C_{10},\delta)}\right]$ is negative for $i \in \mathbb{N}$. $\mathbb{E}\left[\left|\mathcal{N}(0,1)^{\otimes d}\right|^m\right]$ here simply denotes the mth moment of the norm of a d-dimensional standard normally distributed random variable. Therefore it is sufficient to show $\mathbb{E}\left[\xi_i^{(C_{10},\delta)} \middle| \mathcal{F}_{i-1}\right] \leq 0$ for suitable C_{10} and δ. On $\{d_{i-1} < 1\}$ and on $\{r_{i-1} \leq R\}$ this is evident. On $\{d_{i-1} \geq 1, r_{i-1} > R\}$ we use Lemma 4.1.2. Because of 2. there is $0.5 \geq \rho > 0$ with $G''(1, \rho) =: 2\delta > 0$. 1. yields the existence of $C_{10} > 0$ and $\hat{r} > 0$ such that we have for $r > \hat{r}$: $G'(C_{10}, 1, r) < \delta$. Now choose $x, y \in \gamma$ with $\|x_{i-1} - P\| = r_{i-1}$ and $\|y_{i-1} - x_{i-1}\| = \rho$. With 3. we conclude ($\tau \equiv i - 1$ and $z \equiv P$) that for $l_1 := \|x_{i-1} - P\| \wedge \|y_{i-1} - P\|$ and

$l_2 := \|x_i - P\| \wedge \|y_i - P\|$ we have

$$\mathbb{E}\left[(r_i \vee (r_{i-1} - C_{10}) - r_{i-1})\mathbb{1}_{\{r_{i-1} > R, d_{i-1} \geq 1\}} \big| \mathcal{F}_{i-1}\right]$$
$$\leq \mathbb{E}\left[(l_2 \vee (l_1 - C_{10}) - l_1) | \mathcal{F}_{i-1}\right] \mathbb{1}_{\{r_{i-1} > R, d_{i-1} \geq 1\}}$$
$$\leq (G'(C_{10}, 1, r_{i-1}) - G''(1, \rho))\mathbb{1}_{\{r_{i-1} > R, d_{i-1} \geq 1\}} \leq -\delta \mathbb{1}_{\{r_{i-1} > R, d_{i-1} \geq 1\}},$$

provided we choose $R := \hat{r}$ (which we do). So we can apply Lemma 4.1.3 to $(\xi_i^{(\delta, C_{10})} : i \in \mathbb{N})$ for these C_{10} and δ. We will abbreviate $\xi_i^{(C_{10}, \delta)}$ as ξ_i. Fix C_{10} and δ such that the $\xi_i^{(C_{10}, \delta)}$ satisfy the assumptions of Lemma 4.1.3 and conclude for $n \geq 4 \vee 4(r_0 - R)\delta^{-1} \vee (16C_{10} + 1)^2 \delta^{-2}$ that for $m \in \mathbb{N}$ there is $\kappa_m^{(4)} \in \mathbb{R}$ such that

$$IV \leq \mathbb{P}\left[\sum_{i=\lfloor\sqrt{n}\rfloor+1}^{n} \xi_i \geq (n - \lfloor\sqrt{n}\rfloor)\frac{\delta}{4}, \sum_{i=1}^{n} \xi_i \geq \frac{\delta n}{16}\right]$$
$$+ \mathbb{P}\left[\sum_{i=\lfloor\sqrt{n}\rfloor+1}^{n} \xi_i \geq (n - \lfloor\sqrt{n}\rfloor)\frac{\delta}{4}, \sum_{i=1}^{\lfloor\sqrt{n}\rfloor} \xi_i \leq -\frac{\delta n}{16}\right]$$
$$\leq \mathbb{P}\left[\sum_{i=1}^{n} \xi_i \geq \frac{\delta n}{16}\right] + \mathbb{P}\left[\sum_{i=1}^{\lfloor\sqrt{n}\rfloor} \xi_i \leq -\frac{\delta n}{16}\right] \leq \kappa_m^{(4)} n^{-m}. \quad (4.12)$$

Observe that due to $n > (16C_{10})^2 \delta^{-2} \Rightarrow \sum_{i=1}^{\lfloor\sqrt{n}\rfloor} \frac{\xi_i}{\lfloor\sqrt{n}\rfloor} \geq -C_{10} > -\frac{\delta\sqrt{n}}{16}$ the last term vanishes. Combining the equations (4.7), (4.8), (4.9), (4.10) and (4.12) yields for $n \geq 4 \vee 4(r_0 - R)\delta^{-1} \vee (16C_{10} + 1)^2 \delta^{-2}$ that

$$\mathbb{P}\left[\tau^R(\gamma, P) > n\right] \leq \kappa_m^{(6)} n^{-m} + \kappa_m^{(3)} n^{-m} + \kappa_m^{(4)} n^{-m} =: \kappa_m^{(5)} n^{-m},$$

which proves that for $m \in \mathbb{N}$ the choice

$$\kappa_m^{(2)} := \left[(\kappa_m^{(5)} \vee 1) \sup_{r > 1}\left(\frac{r}{\lfloor r \rfloor}\right)^m\right]\left[4 \vee \left(\frac{16C_{10}+1}{\delta}\right)^2 \vee \frac{4}{\delta}\right] < \infty$$

is appropriate as the following computations show.

$$\begin{aligned}
\mathbb{P}\left[\tau^R(\gamma, P) > \kappa_m^{(2)} \beta \bar{r}\right] &= \mathbb{P}\left[\tau^R(\gamma, P) > \left((\kappa_m^{(5)} \vee 1) \sup_{r>1} \left(\frac{r}{\lfloor r \rfloor}\right)^m\right) \left(4 \vee \left(\frac{16C_{10}+1}{\delta}\right)^2 \vee \frac{4}{\delta}\right) \beta \bar{r}\right] \\
&\leq \mathbb{P}\left[\tau^R(\gamma, P) > \left\lfloor \left(4 \vee \left(\frac{16C_{10}+1}{\delta}\right)^2 \vee \frac{4}{\delta}\right) \beta \bar{r} \right\rfloor\right] \\
&\leq \kappa_m^{(5)} \left(\frac{\left(4 \vee \left(\frac{16C_{10}+1}{\delta}\right)^2 \vee \frac{4}{\delta}\right) \beta \bar{r}}{\left\lfloor \left(4 \vee \left(\frac{16C_{10}+1}{\delta}\right)^2 \vee \frac{4}{\delta}\right) \beta \bar{r} \right\rfloor}\right)^m \left(\left[4 \vee \left(\frac{16C_{10}+1}{\delta}\right)^2 \vee \frac{4}{\delta}\right] \beta \bar{r}\right)^{-m} \\
&\leq (\kappa_m^{(5)} \vee 1) \sup_{r>1} \left(\frac{r}{\lfloor r \rfloor}\right)^m (\beta \bar{r})^{-m} \leq \kappa_m^{(2)} \beta^{-m} \bar{r}^{-m}.
\end{aligned}$$

The proof of Theorem 4.3.1 is complete. □

4.3.2 Linear Expansion And Stable Norm

Implications Of Theorem 4.3.1

For collecting the following corollaries of Theorem 4.3.1 we need some notation.

$$\mathcal{W}_t(\gamma) := \bigcup_{0 \leq s \leq t} \gamma_s,$$
$$\mathcal{W}_t^R(\gamma) := \left\{x \in \mathbb{R}^d : \mathrm{dist}\,(x, \mathcal{W}_t(\gamma)) \leq R\right\}.$$

Corollary 4.3.6 (slow linear expansion).
There are positive constants C_{11} and R, such that $\mathbb{P}$-a.s. we have for large t that

$$K_{C_{11}t}(0) \subset \mathcal{W}_t^R(\gamma).$$

$K_r(x)$ *denotes the closed r-Ball centered at x as before.*

Proof: Cover $K_{C_{11}t}(0)$ with balls of radius R. Due to Theorem 4.3.1 the probability, that a predescribed one of these balls has not been hit by γ up to time t, decays faster than any power of t, if we choose C_{11} small enough and R large enough. For the number of balls needed to cover $K_{C_{11}t}(0)$ only grows like t^d the probability that any of these balls has not been hit up to time t decays faster than any power of t provided R is sufficiently large and C_{11} sufficiently small. So Corollary 4.3.6 follows for $t \in \mathbb{N}$ from the first Borel-Cantelli Lemma. Decreasing C_{11} a bit more proves it for general t. $\square$

For the sequel fix $R > 0$ large enough for Theorem 4.3.1 and Corollary 4.3.6 to hold with this R. Assuming that γ is large makes all the estimates of Theorem 4.3.1 uniform in $\gamma \in \mathcal{C}_R$ with $\mathcal{C}_R := \{\gamma : \operatorname{diam}(\gamma) \geq 1, \gamma \subset K_{2R}(0)\}$ (w.l.o.g. we assume $R > 1$). The following corollary is immediate.

Corollary 4.3.7 (uniform integrability)**.**
The family of random variables $\left(\left(\frac{\tau^R(\gamma, tv)}{t}\right)^k\right)_{t \geq 1, \|v\|=1, \gamma \in \mathcal{C}_R}$ is uniformly integrable for any $k \in \mathbb{N}$.

Proof: For $t \geq 1$, $m \in \mathbb{N}$ and $\sqrt[2k]{n} \geq \kappa_m^{(2)}$ we have

$$\mathbb{P}\left[\left(\frac{\tau^R(\gamma, tv)}{t}\right)^k > n\right] \leq \mathbb{P}\left[\tau^R(\gamma, tv) > \kappa_m^{(2)} \sqrt[2k]{nt}\right] \leq \kappa_m^{(2)} (\sqrt[2k]{nt})^{-m} \leq \kappa_m^{(2)} n^{-\frac{m}{2k}},$$

which implies Corollary 4.3.7 for large m because $\kappa_m^{(2)}$ does depend neither on $t \geq 1$, $\|v\| = 1$ nor on $\gamma \in \mathcal{C}_R$. $\square$

The Stable Norm

Set $|v|^R := \sup_{\gamma \in \mathcal{C}_R} \mathbb{E}\left[\tau^R(\gamma, v)\right]$ which due to the isotropic properties of the flow does not depend on the direction of v. We obviously have

$$\mathbb{E}\left[\tau^{2R}(\gamma, (t_1 + t_2)v)\right] \leq \mathbb{E}\left[\tau^R(\gamma, t_1 v)\right] + \sup_{\check{\gamma} \in \mathcal{C}_R} \mathbb{E}\left[\tau^R(\check{\gamma}, t_2 v)\right]. \tag{4.13}$$

With Theorem 4.3.1 we get in addition

$$\mathbb{E}\left[\tau^R(\gamma, (t_1 + t_2)v)\right] \leq \mathbb{E}\left[\tau^{2R}(\gamma, (t_1 + t_2)v)\right] + C_{12} \tag{4.14}$$

for some constant $C_{12} > 0$. We may choose e.g. $C_{12} := \inf_{\gamma \in \mathcal{C}_R, |v| \leq 2R} \mathbb{E}\left[\tau^R(\gamma, v)\right]$. Combining (4.13) and (4.14) yields the subadditivity of $t \mapsto |tv|^R + C_{12}$. Using Feketes lemma we conclude that

$$\|v\|^R := \lim_{t \to \infty} \frac{|tv|^R + C_{12}}{t} = \lim_{t \to \infty} \frac{|tv|^R}{t}$$

is well-defined i.e. the limit exists and equals $\inf_{t\geq 0}(|tv|^R + C_{12})t^{-1}$.
Since $|v|^R$ only depends on $\|v\|$ and since it is increasing with respect to this argument we get (again from the isotropy of the flow) that for $0 \leq s \leq 1$ we have

$$\|sv_1 + (1-s)v_2\|^R \leq s\|v_1\|^R + (1-s)\|v_2\|^R.$$

Set $\mathcal{B} := \{v \in \mathbb{R}^d : \|v\|^R \leq 1\}$ and observe that $\mathcal{B}$ is a compact convex set (see Lemma 4.1.4). Corollary 4.3.6 shows $\|v\|^R \neq 0$ provided $v \neq 0$. Of course the isotropic properties of the flow imply that $\mathcal{B}$ is a ball centered at the origin. We will show later that its radius does not depend on R. First we can prove the following lemma.

Lemma 4.3.8 (lower bound - weak version).
For any $\gamma \in \mathcal{C}_R$ and $\epsilon > 0$ there is $\mathbb{P}$-a.s. $T(\gamma, \epsilon) > 0$, such that we have for $t > T(\gamma, \epsilon)$ that

$$(1-\epsilon)t\mathcal{B} \subset \mathcal{W}_t^R(\gamma).$$

Proof: We need to show that for v with $\|v\|^R \leq 1$ and $m \in \mathbb{N}$ there is $\kappa_m^{(7)} = \kappa_m^{(7)}(\epsilon) > 0$, such that

$$\mathbb{P}\left[\tau^R(\gamma, tv) \geq (1+\epsilon)t\right] \leq \kappa_m^{(7)} t^{-m} \tag{4.15}$$

holds uniformly in $\gamma \in \mathcal{C}_R$ and $\|v\|^R \leq 1$. All the estimates we made so far are uniform in $\|v\|^R = 1$ because they do not depend on the direction of v. By definition of $\|.\|^R$ there is $\tilde{t} > 0$ with $\mathbb{E}\left[\tau^R(\gamma, tv)\right] \leq (1+\frac{\epsilon}{2})t$ for any $t \geq \tilde{t}$ and $\gamma \in \mathcal{C}_R$. Define the stopping time τ_1^R via

$$\tau_1^R := \inf\left\{t > 0, \gamma_t \cap K_R(\tilde{t}v) \neq \emptyset, \operatorname{diam}(\gamma_t) \geq 1\right\}.$$

Denote by $\gamma^{(1)}$ a large connected subset of $\gamma_{\tau_1^R}$ which is contained in $K_{2R}(\tilde{t}v)$ and which has a non-empty intersection with $K_R(\tilde{t}v)$. We can choose it to be $\mathcal{F}_{\tau_1^R}$-measurable which we do. Now define an increasing sequence of stopping times $\left(\tau^{(i)} : i \in \mathbb{N}\right)$ recursively via

$$\tau_i^R := \inf\left\{t > \tau_{i-1}^R, \operatorname{diam}\left(\Phi_{\tau_{i-1}^R, t}\left(\gamma^{(i-1)}\right)\right) \geq 1, \Phi_{\tau_{i-1}^R, t}\left(\gamma^{(i-1)}\right) \cap K_R(i\tilde{t}v) \neq \emptyset\right\},$$
$$\gamma^{(i)} := \Phi_{\tau_{i-1}^R, \tau_i^R}\left(\gamma^{(i-1)}\right) \cap K_{2R}(\tilde{t}iv).$$

If necessary we choose a subset of $\gamma^{(i)}$ as $\gamma^{(i)}$ to ensure that it is connected and has a

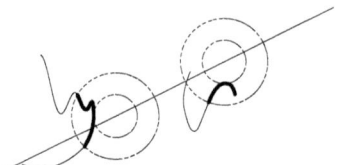

Figure 4.4: line: direction of v, fat: $\gamma^{(i)}$, regular: rest of $\Phi_{\tau_i^R}(\gamma)$

non-empty intersection with $K_R(\tilde{t}iv)$. We have (putting $\tau_0^R \equiv 0$) that

$$\tau^R(\gamma, n\tilde{t}v) \leq \sum_{j=1}^{n}(\tau_j^R - \tau_{j-1}^R). \tag{4.16}$$

Due to the strong Markov-property, the isotropy of Φ and the definition of $\tilde{t}$ the following holds.

$$\mathbb{E}\left[\tau_j^R - \tau_{j-1}^R \big| \mathcal{F}_{\tau_{j-1}^R}\right] \leq \left(1 + \frac{\epsilon}{2}\right)\tilde{t}.$$

Due to Theorem 4.3.1 we can define $\xi_j := \left(\tau_j^R - \tau_{j-1}^R - (1 + \frac{\epsilon}{2})\tilde{t}\right)$ and obtain that the sequence $(\xi_j : j \in \mathbb{N})$ satisfies the assumptions of Lemma 4.1.3. So we conclude

$$\mathbb{P}\left[\tau^R(\gamma, n\tilde{t}v) \geq (1+\epsilon)n\tilde{t}\right] \leq \mathbb{P}\left[\sum_{j=1}^{n}\tau_j^R - \tau_{j-1}^R \geq (1+\epsilon)n\tilde{t}\right]$$

$$= \mathbb{P}\left[\sum_{j=1}^{n}\xi_j \geq \frac{\epsilon}{2}n\tilde{t}\right] \leq \kappa_m^{(1)} n^{-m} = \kappa_m^{(1)} \tilde{t}^m (n\tilde{t})^{-m} =: \kappa_m^{(7)}\left(n\tilde{t}\right)^{-m}$$

which implies that a.s. for any $\epsilon > 0$ the inclusion $(1-\epsilon)n\tilde{t}\mathcal{B} \subset \mathcal{W}_{n\tilde{t}}^R(\gamma)$ fails to hold only a finite number of times. Consider

$$t^{\downarrow} := \left\lfloor \frac{t}{\tilde{t}} \right\rfloor \tilde{t} \Rightarrow \lim_{t \to \infty} \frac{t^{\downarrow}}{t} = 1 \Rightarrow \forall \epsilon > 0 : \exists \check{t} > 0 : \forall t \geq \check{t} : t^{\downarrow} \geq \frac{1-\epsilon}{1-\frac{1}{2}\epsilon}t.$$

Finally we get for $t \geq \check{t} \vee \max\left\{n \in \mathbb{N} : (1 - \frac{\epsilon}{2})n\tilde{t}\mathcal{B} \not\subset \mathcal{W}_{n\tilde{t}}^R(\gamma)\right\}\tilde{t}$ that

$$(1-\epsilon)t\mathcal{B} \subset \left(1 - \frac{\epsilon}{2}\right)t^{\downarrow}\mathcal{B} \subset \mathcal{W}_{t^{\downarrow}}^R \subset \mathcal{W}_t^R \text{ a.s.}$$

and so the proof of Lemma 4.3.8 is complete. □

4.3.3 Sweeping Lemma And A Sharp Lower Bound - The Two-Dimensional Case

In this subsection assume $d = 2$. We can also assume that γ is a curve (which we could have assumed before). In this case we have the following.

Theorem 4.3.9 (lower bound - complete version).
For any $\gamma \in \mathcal{C}_R$ and $\epsilon > 0$ there is $\mathbb{P}$-a.s. $T(\gamma, \epsilon) > 0$, such that for any $t > T(\gamma, \epsilon)$ the following holds.

$$(1 - \epsilon)t\mathcal{B} \subset \mathcal{W}_t(\gamma).$$

This is 1. of Theorem 4.2.1.

Note that we do not distinguish between the $T(\gamma, \epsilon)$ here and the $T(\gamma, \epsilon)$ of Lemma 4.3.8 because the two times are very close to each other as we will see in the sequel. The proof of Theorem 4.3.9 depends apart from Lemma 4.3.8 on the following sweeping lemma, which will be proved after Theorem 4.3.9.

Lemma 4.3.10 (sweeping lemma).
Let γ be a curve with $\mathrm{dist}(\gamma, P) \leq R$ (for an R as defined before). Define

$$\tilde{\tau} = \tilde{\tau}^R(\gamma, P) := \tilde{\tau}(P) := \inf_{t>0}\{K_R(P) \subset \bigcup_{0 \leq s \leq t} \gamma_s\}.$$

Then for $m \in \mathbb{N}$ there is $\kappa_m^{(8)} \in \mathbb{R}$ such that $\mathbb{P}\left[\tilde{\tau} > t\right] \leq \kappa_m^{(8)} t^{-m}$ holds uniformly in γ.

Proof of Theorem 4.3.9: There is a positive integer k, such that for any $n \in \mathbb{N}$ the set $(1 - \epsilon)n\mathcal{B}$ can be covered with $n^2 k$ balls $\{K_R(P_i^n) : i = 1, \ldots, n^2 k\}$ of radius R. By (4.15) the probability, that one of these balls has not been hit by the (at the hitting time large) curve γ up to time $(1 - 0.5\epsilon)n$ decays faster than any power of n. Due to Lemma 4.3.10 $\mathbb{P}\left[\tilde{\tau}(P_i^n) - \tau_R(\gamma, P_i^n) \geq 0, 5\epsilon n\right]$ decays faster than any power of n, too. So the probability that there is one among the balls $K_R(P_i^n)$ for $i \in \{1, \ldots, n^2 k\}$ that is not completely included in $\mathcal{W}_n$ at time n decays faster than any power of t, which proves Theorem 4.3.9 because we have for large t that $(1 - 2\epsilon)t\mathcal{B} \subset (1 - \epsilon)\lfloor t \rfloor \mathcal{B} \subset \mathcal{W}_{\lfloor t \rfloor} \subset \mathcal{W}_t$.

$\square$

Proof of Lemma 4.3.10: The proof consists of six steps. These are carried out similarly to a proof in [13]. This means in detail:

1. localizing of Lemma 4.3.10
2. definition of a small square
3. reduction to the problem of sweeping this in a finite time interval w.p.p.
4. approaching the small square
5. reduction to a control problem
6. construction of a sweeping control

Localizing Of Lemma 4.3.10

Assume we can prove the following. For any $Q \in K_R(P)$ there is an open superset U_Q of Q, such that for any $\tilde{\tau}_Q := \inf_{t>0} \{U_Q \subset \cup_{0 \leq s \leq t} \gamma_s\}$ the following holds. For $m \in \mathbb{N}$ there is $\kappa_m^{(9)} \in \mathbb{R}$, such that

$$\mathbb{P}\left[\tilde{\tau}_Q > t\right] \leq \kappa_m^{(9)} t^{-m} \tag{4.17}$$

holds uniformly for large curves γ which have a non-empty intersection with $K_R(P)$. For the covering of $\overline{K_R(P)}$ requires only a finite number of the U_Q Lemma 4.3.10 holds because of $\{\tilde{\tau} > t\} \subset \{\tilde{\tau}_Q > t$ for one of these $Q\}$.

Definition Of A Small Square

Set $(q_1, q_2) := Q \in K_R(P)$ and consider the following elements of the RKHS $\mathcal{H}$ of Φ.

$$V_1^i(.) := \int b^{ij}(. - y) d\delta_Q \otimes \delta_1(y, j) = b^{i,1}(. - Q) : i = 1, 2;$$
$$V_2^i(.) := \int b^{ij}(. - y) d\delta_Q \otimes \delta_2(y, j) = b^{i,2}(. - Q) : i = 1, 2.$$

We have $w_1 := V_1(Q) = b^{\cdot 1}(0) = \begin{pmatrix} 1 \\ 0 \end{pmatrix}$ and $w_2 := V_2(Q) = b^{\cdot 2}(0) = \begin{pmatrix} 0 \\ 1 \end{pmatrix}$. Due to Lemma 1.3.2 the following holds for $\tilde{x} := x - Q$.

$$(V_1(\tilde{x}), V_2(\tilde{x})) = E_2 + O\left(\|\tilde{x}\|^2\right) : (\tilde{x} \to 0)$$

4.3 THE LOWER BOUND

where the last equation is to be understood in components. So there are $C_{13} > 0$ and $\delta > 0$ such that we have for $\|x - Q\| < \delta$ that

$$\|V_1(x) - w_1\| \vee \|V_2(x) - w_2\| \leq C_{13} \|x - Q\|^2.$$

This implies that for $n \in \mathbb{N}$ there is $\epsilon > 0$ such that for

$$U_Q^{n,\epsilon} := \,]q_1 - n\epsilon, q_1 + n\epsilon[\, \times \,]q_2 - n\epsilon, q_2 + n\epsilon[$$

we have that

$$U_Q^{n,\epsilon} \subset \left\{ y \in \mathbb{R}^2 : \|V_1(y) - w_1\| \vee \|V_2(y) - w_2\| \leq \epsilon \right\},$$

because for $\epsilon \leq 2^{-0.5} n^{-1} \delta \wedge (2C_{13} n^2)^{-1}$ and $x := (x_1, x_2) \in U_Q^{n,\epsilon}$ we get

$$\|V_1(x) - w_1\| \vee \|V_2(x) - w_2\| \leq C_{13} \left[(x_1 - q_1)^2 + (x_2 - q_2)^2 \right] \leq 2C_{13} n^2 \epsilon^2 \leq \epsilon.$$

Note that this still holds if we decrease ϵ (for a fixed n). Define

$$\tilde{U}_Q^n := \,]q_1 - \frac{n\epsilon}{2}, q_1 + \frac{n\epsilon}{2}\Big[\times\Big]q_2 - \frac{n\epsilon}{2}, q_2 + \frac{n\epsilon}{2}\Big[\text{ and } t_u^n := \frac{n\epsilon}{2} \left(\sup_{z \in U_Q^{n,\epsilon}} (\|V_1(z)\| \vee \|V_2(z)\|) \right)^{-1} > 0.$$

We may assume $t_u^n \geq 3^{-1} n\epsilon$ as well as $\epsilon \leq 102^{-1}$ (and otherwise choose a smaller ϵ). Denote by $\psi_{st}^{(i)}(x)$ for $i = 1, 2$ the deterministic flow defined to be the solution of the control problem

$$\psi_{st}^{(i)}(x) = x + \int_s^t V_i \left(\psi_{sr}^{(i)}(x) \right) dr...$$

Proposition 4.3.11 (nice behavior of ψ).
For $t \leq t_u^n$, $z \in \tilde{U}_Q^n$ and $i = 1, 2$ we have $\left\| \psi_{0,t}^{(i)}(z) - z - tw_i \right\| \leq \epsilon t$.

Proof of Proposition 4.3.11: For $z \in \tilde{U}_Q^n$ and $i = 1, 2$ we have $\|V_i(z)\| \leq 0,5n\epsilon (t_u^n)^{-1}$, which implies for $z \in \tilde{U}_Q^n$ and $i = 1, 2$ that

$$\inf_{t>0} \left\{ \left\| \psi_{0t}^{(i)}(z) - z \right\| > \frac{n\epsilon}{2} \right\} \geq t_u^n \Rightarrow \inf_{t>0} \left\{ \psi_{0t}^{(i)}(z) \notin U_Q^n \right\} \geq t_u^n,$$

which proves Proposition 4.3.11 because we have for $z \in \tilde{U}_Q^n$ that

$$\left\|\psi_{0t}^{(i)}(z) - z - tw_i\right\| \leq \int_0^t \left\|V_i\left(\psi_{0s}^{(i)}(z)\right) - w_i\right\| ds \leq \epsilon t.$$

$\square$

Now we consider the coordinates $Z := (Z_1, Z_2) : \tilde{U}_Q^{102} \to]-51, 51[$ generated by the constant vector fields $(\epsilon w_i : i = 1, 2)$ and choose $U_Q := Z^{-1}(]-1, 1[^2)$.

From Large To Positive Probability

As we will see it suffices to show that there is $0 < \theta < 1$ such that for a large curve with a non-empty intersection with $K_R(P)$ we have uniformly in $Q \in K_R(P)$ that

$$\mathbb{P}\left[\tilde{\tau}_Q < t_j \,|\, \tilde{\tau}_Q > t_{j-1}\right] \geq \theta. \tag{4.18}$$

Therein for a $T > 0$ (to be specified later) let $t_0 := 0$ and for $j \in \mathbb{N}$

$$t_j := \inf\left\{t \in \mathbb{R} : t \geq t_{j-1} + 1 + T : \gamma_t \cap K_R(P) \neq \emptyset, \operatorname{diam}(\gamma_t) \geq 1\right\}.$$

Following Theorem 4.3.1 there is $C_{14} > 0$ and for $m \in \mathbb{N}$ a $\kappa_m^{(10)} \in \mathbb{R}$, such that for $j \in \mathbb{N}$ we have $\mathbb{P}[t_j > C_{14}j] \leq \kappa_m^{(10)} j^{-m}$ which implies

$$\mathbb{P}\left[\tilde{\tau}_Q > t\right] \leq \mathbb{P}\left[t_{\lfloor \frac{t}{C_{14}} \rfloor} > t\right] + \mathbb{P}\left[\tilde{\tau}_Q > t_{\lfloor \frac{t}{C_{14}} \rfloor}\right]$$

$$\leq \mathbb{P}\left[t_{\lfloor \frac{t}{C_{14}} \rfloor} > C_{14}\left\lfloor \frac{t}{C_{14}} \right\rfloor\right] + \mathbb{P}\left[\tilde{\tau}_Q > t_{\lfloor \frac{t}{C_{14}} \rfloor} \,\Big|\, \tilde{\tau}_Q > t_{\lfloor \frac{t}{C_{14}} \rfloor - 1}\right] \mathbb{P}\left[\tilde{\tau}_Q > t_{\lfloor \frac{t}{C_{14}} \rfloor - 1}\right]$$

$$\leq \mathbb{P}\left[t_{\lfloor \frac{t}{C_{14}} \rfloor} > C_{14}\left\lfloor \frac{t}{C_{14}} \right\rfloor\right] + (1-\theta)\mathbb{P}\left[\tilde{\tau}_Q > t_{\lfloor \frac{t}{C_{14}} \rfloor - 1} \,\Big|\, \tilde{\tau}_Q > t_{\lfloor \frac{t}{C_{14}} \rfloor - 2}\right] \mathbb{P}\left[\tilde{\tau}_Q > t_{\lfloor \frac{t}{C_{14}} \rfloor - 2}\right]$$

$$\leq \mathbb{P}\left[t_{\lfloor \frac{t}{C_{14}} \rfloor} > C_{14}\left\lfloor \frac{t}{C_{14}} \right\rfloor\right] + \ldots \leq \kappa_m^{(10)}\left\lfloor \frac{t}{C_{14}} \right\rfloor^{-m} + (1-\theta)^{\lfloor \frac{t}{C_{14}} \rfloor} \leq \kappa_m^{(9)} t^{-m}$$

for suitable $\kappa_m^{(9)} \in \mathbb{R}$. So we have only to prove (4.18). Therefore it is enough to show that for γ (as before) there are $T > 0$ and $\theta > 0$ (not depending on the chosen γ) such that we

4.3 THE LOWER BOUND

have uniformly in $Q \in K_R(P)$ that

$$\mathbb{P}\left[U_Q \subset \bigcup_{0\leq s \leq T} \gamma_s\right] \geq \theta.$$

Approaching The Small Square

Let $\hat{U}_Q := Z^{-1}(]-7,7[^2)$. Then we have obviously $U_Q \subset \hat{U}_Q$. Choose x and y in γ with $\|x - P\| \leq R$ and $\|x - y\| \geq 0.5$. Due to the Lemmas 1.3.7 and 1.3.2 the eigenvalues of the matrix $\bar{b}^2(z) := \bar{b}^*(z)\bar{b}(z)$ are bounded below by a positive constant C_4 on $\{\|z\| \geq \delta\}$ for arbitrary $\delta > 0$. The boundedness of the correlation functions gives an upper bound C_5. Therefore the $\mathbb{R}^4$-valued semimartingale

$$\left(\begin{pmatrix} x_t - (x + 2t(Q-x)) \\ y_t - (y + 2t(Q-x)) \end{pmatrix} : t \in \left[0, \frac{1}{2}\right]\right)$$

satisfies the assumptions of Lemma 3.2.1. So this lemma yields for $t = 0.5$ and $\delta = 0.5\epsilon$ (C_4, C_5 and C_6 can be chosen to be independent of x and y) that

$$\mathbb{P}\left[x_{\frac{1}{2}} \in U_Q, y_{\frac{1}{2}} \notin \hat{U}_Q\right] \geq p$$

where $p > 0$ does not depend on the special choice of γ, because $\epsilon \leq (56\sqrt{2})^{-1}$ implies that $\operatorname{diam}(\hat{U}_Q) \leq 0.25$. Denote by $\hat{\hat{\gamma}}$ the subcurve of γ, between $x_{0.5}$ and $y_{0.5}$ and by $\hat{\gamma}$ a minimal subcurve of $\hat{\hat{\gamma}}$, which is contained in $\hat{U}_Q$ and which links $\partial \hat{U}_Q$ to ∂U_Q (minimal means that no proper subcurve has these properties). Due to minimality of $\hat{\gamma}$ the set $\hat{\gamma} \cap \partial \hat{U}_Q$ consists of a single point which we will denote by z. $\partial \hat{U}_Q$ consists of four pieces. Without loss of generality assume $z \in Z^{-1}(\{-7\} \times [-7,7])$ (the other cases are similar). Let $\tilde{\gamma}$ be the minimal subcurve of $\hat{\gamma}$, linking z with $Z^{-1}(\{-1\} \times [-7,7])$ and $\tilde{y} := \tilde{\gamma}_{0.5} \cap Z^{-1}(\{-1\} \times [-7,7])$ the intersection point (see Fig. 4.5). We have to show that there exist $T > 0$ and $\theta > 0$ such that for any curve $\tilde{\gamma} \subset Z^{-1}([-7,7] \times [-7,7])$ linking $Z^{-1}(\{-7\} \times [-7,7])$ to $Z^{-1}(\{-1\} \times [-7,7])$ the following holds.

$$\mathbb{P}\left[U_Q \subset \bigcup_{0 \leq s \leq T} \tilde{\gamma}_s\right] \geq \theta. \tag{4.19}$$

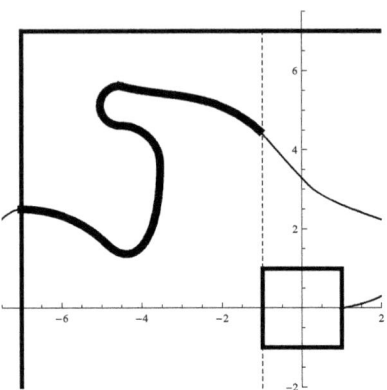

Figure 4.5: $\tilde{\gamma}$ (fat), rest of $\hat{\gamma}$ (regular), the endpoints of $\tilde{\gamma}$ are z (left) and $\tilde{y}$ (right).

Reduction To A Control Problem

Definition 4.3.12 ($\mathcal{H}$-simple control).
Denote by $\mathcal{H}$ the RKHS of Φ. An $\mathcal{H}$-simple control V is a mapping from $[0,T]$ to $\mathcal{H}$, which is piecewise constant.

For an $\mathcal{H}$-simple control V denote by $\psi_s^{(V)}(x)$ the solution of the control problem

$$\begin{cases} \partial_t \psi_t^{(V)}(x) &= V\left(\psi_t^{(V)}(x)\right) \\ \psi_0^{(V)}(x) &= x \end{cases}.$$

Assume we can construct an $\mathcal{H}$-simple control V with the following property. If $\Psi(.,.)$ is a continuous mapping from $[0,T] \times \mathbb{R}^2$ to $\mathbb{R}^2$ which satisfies

$$\left| Z\left(\Psi(s,x)\right) - Z\left(\psi_s^{(V)}(x)\right) \right| < \frac{1}{2} \tag{4.20}$$

for $x \in \tilde{\gamma}$ and $s \in [0,T]$, then we have

$$U_Q \subset \bigcup_{x \in \tilde{\gamma}} \bigcup_{0 \leq s \leq 1} \Psi(s,x).$$

Then Theorem 1.3.12 applied to the intervals of constance of V and the independence

4.3 THE LOWER BOUND

properties of Brownian flows prove (4.19). Let $\tilde{\gamma} = (\tilde{\gamma}(u) : u \in [0,1])$ be a parametrization of $\tilde{\gamma}$ and set
$\tilde{\psi}(s,u) := \psi_s^{(V)}(\tilde{\gamma}(u))$ as well as $\tilde{\Psi}(s,u) := \Psi(s,\tilde{\gamma}(u))$. ($\bar{\psi}$ and $\bar{\Psi}$ are defined similarly with $\tilde{\gamma}$ replaced by $\bar{\gamma}$, see its definition below.) We want to construct a $\mathcal{H}$-simple control V that implies for any Ψ fulfilling (4.20) that $U_Q \subset \cup_{\frac{7}{17}T \leq s \leq T} \cup_{0 \leq u \leq 1} \tilde{\Psi}(s,u)$. Set $\bar{\gamma} := \partial\left(\left[\frac{7}{17}T, T\right] \times [0,1]\right)$. V is supposed to yield for $\tilde{Q} \in U_Q$ the following.

$$\text{ind}(\bar{\Psi}, \tilde{Q}) = 1. \tag{4.21}$$

Therein we denote by $\text{ind}(\bar{\Psi}, \tilde{Q})$ the curving number of $\bar{\Psi}$ around $\tilde{Q}$. To show (4.21) for all $\tilde{\Psi}$ with

$$\left\|Z(\tilde{\Psi}(.,.)) - Z(\tilde{\psi}(.,.))\right\|_\infty \leq 0.5 \tag{4.22}$$

we construct V in a way, that provides for $\tilde{Q} \in U_Q$ the following

$$\text{ind}(\bar{\psi}, \tilde{Q}) = 1 \text{ and dist}\left(Z\left(\bar{\psi}\right), Z(U_Q)\right) \geq 1. \tag{4.23}$$

Note that (4.23) implies that for any $\tilde{\Psi}$ satisfying (4.22) we indeed get (4.21). In fact $\tilde{\psi}$ sweeps the entire set $Z^{-1}\left([-2,2]^2\right)$.

Construction Of A Sweeping Control

Consider the $\mathcal{H}$-simple control $V : [0,T] := [0, 34\epsilon] \to \mathcal{H}$,

$$V(.,t) := \begin{cases} -V_2(.) & : t \in [0, 10\epsilon[\\ V_1(.) & : t \in [10\epsilon, 14\epsilon[\\ V_2(.) & : t \in [14\epsilon, 34\epsilon] \end{cases}.$$

This control satisfies all our wishes (see Fig. 4.6). Note that $34\epsilon = \frac{102}{3}\epsilon \leq t_u^{102}$ ensures the suitability of Proposition 4.3.11 for $t \leq 34\epsilon$. So we get for $t \leq 34\epsilon$, $z \in \hat{U}_Q$ and $i = 1, 2$ that

$$\left\|\Psi_{0t}^{(i)}(z) - z - tw_i\right\| \leq 34\epsilon^2 \leq \frac{\epsilon}{3},$$

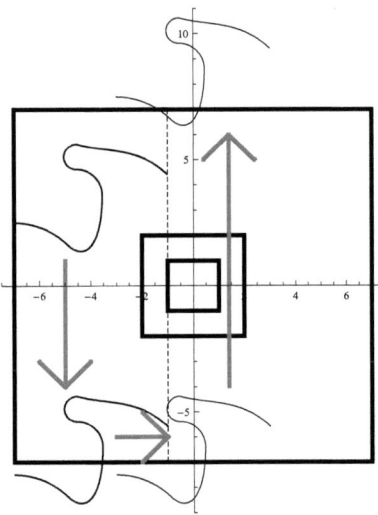

Figure 4.6: γ (curve), ∂U_Q and $Z^{-1}\left(\{-1\} \times [-7,7]\right)$ (dashed) , $Z^{-1}([-2,2]^2)$ and $\partial \hat{U}_Q$ (fat)

4.3 THE LOWER BOUND

which gives the following estimates for $x \in \tilde{\gamma}$.

$$
\begin{array}{llllll}
-7 \leq & Z_1(x) \leq & -1; & -7 & \leq Z_2(x) & \leq 7 & :t = 0, \\
-7 - \frac{1}{3} \leq & Z_1(x_t) \leq & -1 + \frac{1}{3}; & -17 - \frac{1}{3} & \leq Z_2(x_t) & \leq -3 + \frac{1}{3} & :t = 10\epsilon, \\
-3 - \frac{2}{3} \leq & Z_1(x_t) \leq & 3 + \frac{2}{3}; & -17 - \frac{2}{3} & \leq Z_2(x_t) & \leq -3 + \frac{2}{3} & :t = 14\epsilon, \\
-3 - 1 \leq & Z_1(x_t) \leq & 3 + 1; & 3 - 1 & \leq Z_2(x_t) & \leq 17 + 1 & :t = 34\epsilon.
\end{array}
$$

Herein we used that elapsing of time through the intervals of constance of V changes any coordinate as if Z was generated by ϵV with an error of at most 3^{-1}.(see Fig. 4.6). A similar arguing holds for z and $\tilde{y}$.

$$
\begin{array}{llllll}
& Z_1(z) = & -7; & & Z_1(\tilde{y}) = & -1: & t = 0, \\
-7 - \frac{1}{3} \leq & Z_1(z_t) \leq & -7 + \frac{1}{3}; & -1 - \frac{1}{3} \leq & Z_1(\tilde{y}_t) \leq & -1 + \frac{1}{3}: & t = 10\epsilon, \\
-3 - \frac{2}{3} \leq & Z_1(z_t) \leq & -3 + \frac{2}{3}; & 3 - \frac{2}{3} \leq & Z_1(\tilde{y}_t) \leq & 3 + \frac{2}{3}: & t = 14\epsilon, \\
-3 - 1 \leq & Z_1(z_t) \leq & -3 + 1; & 3 - 1 \leq & Z_1(\tilde{y}_t) \leq & 3 + 1: & t = 34\epsilon.
\end{array}
$$

This means that for $t \in [14\epsilon, 34\epsilon]$ z_t is on the left and $\tilde{y}_t$ is on the right of $\check{U}$ which implies (4.23) and completes the proof of Lemma 4.3.10. □

4.3.4 Dependence Of $\|v\|^R$ On R

We will now get rid of the unnatural dependence of $\|\cdot\|^R$ on R.

Lemma 4.3.13 (R does not play any role).
If we define for $R > 0$, $\tilde{R} \geq 1$ and $v \in \mathbb{R}^d$ $\|v\|_{\tilde{R}}^R$ via

$$\|v\|_{\tilde{R}}^R := \lim_{t \to \infty} \frac{\sup_{\gamma \in \mathcal{C}_{\tilde{R}}} \mathbb{E}\left[\tau^R(\gamma, tv)\right]}{t}$$

then this limit exists and we have for arbitrary $R_1 > 0$, $R_2 > 0$ and $\tilde{R}_1 \geq 1$, $\tilde{R}_2 \geq 1$ that

$$\|\cdot\|_{\tilde{R}_1}^{R_1} \equiv \|\cdot\|_{\tilde{R}_2}^{R_2},$$

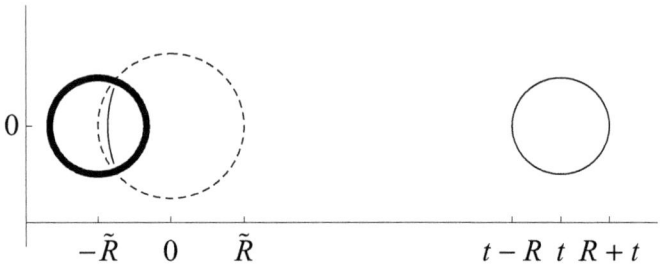

Figure 4.7: $K_R(-\tilde{R}v)$ (black), $K_{\tilde{R}}(0)$ (dashed), $K_R(tv)$ (black), candidate for a slow curve (fat)

i.e. especially $\|v\|^R$ does not depend on R.

Proof: Define for $t \geq 0$, $R > 0$ and $\tilde{R} \geq 1$ the function $\bar{g} = \bar{g}(R, \tilde{R}, t)$ via

$$\bar{g}(R, \tilde{R}, t) := \sup_{\gamma \in \mathcal{C}_{\tilde{R}}} \left\{ \mathbb{E}\left[\tau^R(\gamma, tv)\right] \right\}.$$

Herein fix $v \in \mathbb{R}^d$ with $\|v\| = 1$. As already seen there is $R > 0$ such that

$$\|v\|_R^R := \lim_{t \to \infty} \frac{g(R, R, t)}{t} \tag{4.24}$$

exists (the limit was named $\|v\|^R$). The rest of the proof consists of the application of the following two propositions.

Proposition 4.3.14 (lower R).
Fix $R \geq 1$ in a way that we have convergence in (4.24). Then we have for arbitrary $\tilde{R} \geq 1$ that $\|v\|_{\tilde{R}}^R = \lim_{t \to \infty} \frac{g(R, \tilde{R}, t)}{t} = \|v\|_R^R$.

Proof:

1. If $\tilde{R} \geq R$ then $\bar{g}(R, \tilde{R}, t) \geq \bar{g}(R, R, t)$ is obvious. The isotropy of the flow yields that $\bar{g}(R, \tilde{R}, t) \leq \bar{g}(R, R, t + \tilde{R})$ (see Fig. 4.7). For we have $\mathbb{E}\left[\tau^R\left(\gamma, (t + \tilde{R})v\right)\right] \leq \mathbb{E}\left[\tau^R(\gamma, tv)\right] + C_{15}$ for a $C_{15} > 0$ (uniformly chosen in γ) we get $\bar{g}(R, R, t) \leq \bar{g}(R, \tilde{R}, t) \leq \bar{g}(R, R, t) + C_{15}$.

2. If $\tilde{R} < R$ we obtain similarly that $\bar{g}(R, \tilde{R}, t) \leq \bar{g}(R, R, t) \leq \bar{g}(R, \tilde{R}, t - R) \leq \bar{g}(R, \tilde{R}, t) + C_{15}$.

Sending $t \to \infty$ proves Proposition 4.3.14 from the last equations. □

Proposition 4.3.15 (upper R).
$\|v\|_1^R$ exists for any $R > 0$ and we have $\|.\|_1^R \equiv \|.\|_1^{\tilde{R}}$.

Proof: Without loss of generality assume that $\tilde{R} > R$. Then $\bar{g}(\tilde{R}, 1, t) \leq \bar{g}(R, 1, t)$ is obvious. Additionally (in $\tau^R(\gamma, tv)$ one takes a subcurve if necessary)

$$\mathbb{E}\left[\tau^R(\gamma, tv)\right] \leq \mathbb{E}\left[\tau^{\tilde{R}}(\gamma, tv)\right] + \underbrace{\sup_{\check{\gamma} \in \mathcal{C}_1} \mathbb{E}\left[\tau^R(\check{\gamma}, \tilde{R}v)\right]}_{<\infty \text{ for } R>0},$$

so Proposition 4.3.15 follows via $t \to \infty$ from $\bar{g}(R, 1, t) \leq \bar{g}(\tilde{R}, 1, t) + C_{16}$ for some $C_{16} > 0$. □

The proof of Lemma 4.3.13 is complete because of $\|.\|_{R_2}^{R_1} = \|.\|_1^{R_1} = \|.\|_1^{\tilde{R}_1} = \|.\|_{\tilde{R}_2}^{\tilde{R}_1}$. □

4.4 The Upper Bound

4.4.1 The Speed Of A Slow Curve Asymptotically Has A Dirac Distribution On The Proper Time Scale

Let R be as before and $v \in \mathbb{R}^d$ with $\|v\|^R = 1$. Choose for any $t \in]0, \infty[$ a curve $\gamma^{(t)} \in \mathcal{C}_{R/2}$ such that $|\mathbb{E}\left[\tau^R(\gamma^{(t)}, tv)\right] - |tv|^R|t^{-1} \to 0$. By the definition of $\|v\|^R$ we already know that

$$\left|\frac{\mathbb{E}\left[\tau^R(\gamma^{(t)}, tv)\right]}{t} - \|v\|^R\right| \leq \underbrace{\left|\frac{\mathbb{E}\left[\tau^R(\gamma^{(t)}, tv)\right] - |tv|^R}{t}\right|}_{\leq t^{-1}e^{-t}} + \underbrace{\left|\frac{|tv|^R}{t} - \|v\|^R\right|}_{\to 0}. \quad (4.25)$$

We will investigate the asymptotic law of $\tau^R(\gamma^{(t)}, tv)$ with the following lemma.

Lemma 4.4.1 (Dirac distribution).
Let $(X_t : t > 0)$ be a family of integrable random variables such that we have for any $\delta > 0$ that
$$\lim_{t \to \infty} \mathbb{E}\left[X_t - \mathbb{E}\left[X_t\right] ; X_t - \mathbb{E}\left[X_t\right] > \delta\right] = 0.$$
Then we also have for $\delta > 0$ that $\lim_{t \to \infty} \mathbb{P}\left[X_t - \mathbb{E}\left[X_t\right] < -\delta\right] = 0$.

Proof: Denote $X_t - \mathbb{E}\left[X_t\right]$ by Y_t and suppose that we can find $\delta > 0$, $\epsilon > 0$ and a sequence $(t_n)_{n \in \mathbb{N}}$ such that for $t_n \nearrow \infty$ and any n we have $\mathbb{P}\left[Y_{t_n} < -\delta\right] > \epsilon$. This immediately yields

$$0 = \mathbb{E}\left[Y_{t_n}\right] = \underbrace{\mathbb{E}\left[Y_{t_n} ; Y_{t_n} \leq -\delta\right]}_{\leq -\epsilon\delta} + \underbrace{\mathbb{E}\left[Y_{t_n} ; Y_{t_n} \in \left(-\delta, \frac{\epsilon\delta}{2}\right)\right]}_{\leq \frac{\epsilon\delta}{2}} + \underbrace{\mathbb{E}\left[Y_{t_n} ; Y_{t_n} \geq \frac{\epsilon\delta}{2}\right]}_{\to 0} \quad (4.26)$$

i.e. a contradiction for large n. □

Lemma 4.4.2 (expectation if τ^R is large).
Let R be as before and fix v with $\|v\|^R = 1$. Then for any $\delta > 0, n, m \in \mathbb{N}$ there are $K_m^{(n)}(\delta) \in \mathbb{R}$ and $\tilde{t}_m^{(n)}(\delta)$ such that we have for all $t \geq 0$ that

$$\mathbb{E}\left[\left(\tau^R(\gamma^{(t)}, tv) t^{-1}\right)^n ; \tau^R(\gamma^{(t)}, tv) t^{-1} > (1+\delta)\right] \leq K_m^{(n)}(\delta) t^{-m}$$

and for $t \geq \tilde{t}_m^{(n)}(\delta)$ that

$$\mathbb{E}\left[\left(\tau^R(\gamma^{(t)}, tv) t^{-1}\right)^n ; \tau^R(\gamma^{(t)}, tv) t^{-1} > \mathbb{E}\left[\tau^R(\gamma^{(t)}, tv) t^{-1}\right] + \delta\right] \leq K_m^{(n)}(\delta) t^{-m}.$$

Proof: The first inequality follows from straightforward estimates using (4.15) and the second one is implied by the first one and (4.25). □

The previous lemma implies that we can apply Lemma 4.4.1 to $X_t := \tau^R(\gamma^{(t)}, tv) t^{-1}$ to conclude that X_t converges to 1 in probability.

4.4.2 Time Reverse - Comparison Of Fast And Slow Curves

We will have to assume $d = 2$ from now on (unless otherwise stated) for the following arguments strongly depend on the topology of the plane.

4.4 THE UPPER BOUND

Theorem 4.4.3 (uniform speed for all curves).
Let $\Gamma := \partial K_R(0)$. There are $C_{20} > 0$ and $C_{17} > 0$ and for $m \in \mathbb{N}$ $\kappa_m^{(11)} \in \mathbb{R}$ such that we have for $T \geq \sqrt{t}$ and any $\gamma \in \mathcal{C}_{R/2}$ that

$$\kappa_m^{(11)} T^{-m} + \mathbb{P}\left[\tau^R(\gamma, tv) \leq T + C_{17}\right] \geq C_{20} \mathbb{P}\left[\tau^R(\Gamma, tv) \leq T\right].$$

The proof of Theorem 4.4.3 uses the following lemmas. Denote by $\mathcal{C}_R^*$ the set of all large curves γ with $\gamma \cap \partial K_R(0) \neq \emptyset$.

Lemma 4.4.4 (p_1).
There is a constant $C_{18} > 0$ with $\inf_{\gamma \in \mathcal{C}_R^*} \inf_{t \geq C_{18}} \mathbb{P}\left[\gamma_t \cap \partial K_R(0) \neq \emptyset; \operatorname{diam}(\gamma_t) \geq 1\right] =: p_1 > 0$.

Proof: Let for some $\delta > 0$ $X_t := \left\lfloor \frac{\operatorname{dist}(0, \gamma_t)}{\delta} \right\rfloor$ and define the sequence of stopping times $(\tau_n : n \in \mathbb{N})$ via $\tau_0 := 0$ and $\tau_n := \inf\{t \geq \tau_n : X_t \neq X_{\tau_n}\}$. By the continuity of $t \mapsto \frac{\operatorname{dist}(0, \gamma_t)}{\delta}$ we get that $\mathbb{P}\left[|X_{\tau_{n+1}} - X_{\tau_n}| = 1 \,|\, \tau_{n+1} < \infty\right] = 1$. The results of Section 4.3 show that

$$\mathbb{P}\left[\limsup_{k \to \infty} X_{\tau_k} \leq 1 \,\bigg|\, \tau_n = \infty \text{ for some n}\right] = 1$$

provided that the event $\{\tau_n = \infty \text{ for some n}\}$ has positive probability. The reasoning leading to [53, (15)] shows that there exists $N \in \mathbb{N}$ such that for $n \geq N$ and $k \in \mathbb{N}$ we have for $\check{F} := \cap_{n \in \mathbb{N}} \{\operatorname{diam} \gamma_n \geq 1\}$ that

$$1 - \mathbb{P}\left[X_{\tau_{k+1}} = n+1 \,\Big|\, X_{\tau_k} = n, \check{F}\right] = \mathbb{P}\left[X_{\tau_{k+1}} = n-1 \,\Big|\, X_{\tau_k} = n, \check{F}\right] \geq \tilde{p}_1 > \frac{1}{2}$$

Hence on $\check{F}$ for any constant $C_{18} \in \mathbb{N}$ the distance process $(\operatorname{dist}(\gamma_{n-0.5C_{18}}, 0) : n \in \mathbb{N})$ of a large curve from the origin is stochastically dominated by a stationary process $(Y_n : n \in \mathbb{N})$ (unsymmetric random walk on $\mathbb{N}$ with finitely many changes in the transition matrix) there is $\epsilon > 0$ with

$$\inf_{\gamma \in \mathcal{C}_R^*} \inf_{\mathbb{N} \ni t \geq C_{18}} \mathbb{P}\left[\gamma_t \cap K_R(0) \neq \emptyset\right] \geq \inf_{\gamma \in \mathcal{C}_R^*} \inf_{\mathbb{N} \ni t \geq C_{18}} \mathbb{P}\left[\gamma_t \cap K_R(0) \neq \emptyset \,\Big|\, \check{F}\right] \mathbb{P}\left[\check{F}\right]$$

$$\geq \inf_{\gamma \in \mathcal{C}_R^*} \inf_{\mathbb{N} \ni t \geq C_{18}} \mathbb{P}\left[Y_t \leq R \,\Big|\, \check{F}\right] \inf_{\gamma \in \mathcal{C}_R^*} \mathbb{P}\left[\check{F}\right] \geq \epsilon.$$

Now choose C_{18} large enough for $\inf_{\gamma \in \mathcal{C}_R^*} \inf_{t \geq C_{18}} \mathbb{P}\left[\operatorname{diam}(\gamma_t) > 3R\right] \geq 1 - \frac{\epsilon}{2}$ to hold (this

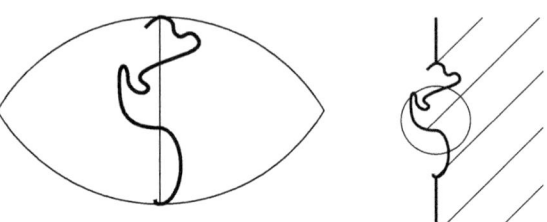

Figure 4.8: Definition of γ_{t_1} via intersection of circles around its endpoints

Figure 4.9: Adding to half lines to γ_{t_1} divides $\mathbb{R}^2$ into black and white parts

probability converges to one cf. (4.6)).
This implies $\inf_{\gamma \in \mathcal{C}_R^*} \inf_{t \geq C_{18}} \mathbb{P}\left[\gamma_t \cap K_R(0) \neq \emptyset; \gamma_t \cap K_R(0)^C \neq \emptyset; \text{diam}(\gamma_t) \geq 1\right] \geq \frac{\varepsilon}{2}$. The passage from $C_{18} \leq t \in \mathbb{N}$ to $C_{18} \leq t \in \mathbb{R}$ is obvious from the Markov Property of the flow and implies $p_1 > 0$. □

Lemma 4.4.5 (p_2).
There is $C_{19} > 0$ such that $\inf_{\bar{\gamma} \in \mathcal{C}_R^*} \inf_{\gamma \in \mathcal{C}_{R/2}} \mathbb{P}\left[\bar{\gamma}_{C_{19}} \cap \gamma \neq \emptyset\right] =: p_2 > 0$.

Proof: First write $t = t_1 + t_2$ for some non-negative t_1, t_2 and observe

$$\mathbb{P}\left[\gamma_t \cap \bar{\gamma} = \emptyset\right] = \mathbb{P}\left[\Phi_{t_1,t}\left(\Phi_{0,t_1}(\gamma)\right) \cap \bar{\gamma} = \emptyset\right] = \mathbb{P}\left[\Phi_{0,t_1}(\gamma) \cap \Phi_{t,t_1}(\bar{\gamma}) = \emptyset\right]$$
$$= (\mathbb{P} \otimes \mathbb{P}) \circ \pi_1^{-1}\left[\Phi_{0,t_1}(\gamma) \cap \Phi_{t,t_1}(\bar{\gamma}) = \emptyset\right] = \mathbb{P} \otimes \mathbb{P}\left[\Phi_{0,t_1}(\gamma) \cap \tilde{\Phi}_{t,t_1}(\bar{\gamma}) = \emptyset\right]$$
$$= \mathbb{P} \otimes \mathbb{P}\left[\Phi_{0,t_1}(\gamma) \cap \tilde{\Phi}_{t_1,t}(\bar{\gamma}) = \emptyset\right] = \mathbb{P} \otimes \mathbb{P}\left[\Phi_{0,t_1}(\gamma) \cap \tilde{\Phi}_{0,t_2}(\bar{\gamma}) = \emptyset\right].$$

Herein we denote by $\tilde{\Phi}$ an independent copy of Φ (e.g. defined on $(\Omega \times \Omega, \mathcal{F} \otimes \mathcal{F}, \mathbb{P} \otimes \mathbb{P})$). So we can (instead of having $\bar{\gamma}$ running t) split t and let γ run one part of the time and $\bar{\gamma}$ the rest of it. We already know that for sufficiently large t_1 γ_{t_1} is dense as much as $\sqrt{t_1}$ in $K_{t_1^{0.9}}(0)$ with probability say at least 0.5 uniformly in γ (this probability converges to 1 as $t_1 \to \infty$ cf [53]). In this case rename γ_{t_1} to be a connected subcurve of $\gamma_{t_1} \cap K_{t_1^{0.9}}(0)$ of diameter $t_1^{0.8}$ that has distance not more than $\sqrt{t_1}$ from the origin. We may asume that the endpoints of the new γ_{t_1} have distance $t_1^{0.8}$ from each other (which we do). Note that γ_{t_1} is contained in the intersection of the $t_1^{0.8}$-balls around its endpoints (see Fig. 4.8). If we add two half-lines to γ_{t_1} then it separates the plane into two parts - say the black part

4.4 THE UPPER BOUND

and the white part (cf. Fig 4.9). Now we fix a ball $\mathcal{K}$ of radius of $t_1^{0.6}$ centered on the perpendicular bisector of the endpoints of γ_{t_1} exactly one half of which is black (measured with Lebesgue measure). The existence of such a ball follows from a continuity argument and the fact that there are completely black and completely white balls centered there. Fix t_1 large enough for $t_1^{0.9} \gg t_1^{0.8} \gg t_1^{0.7} \gg t_1^{0.6} \gg t_1^{0.5} \gg 1$ to hold. Observe now that any curve that links the black part and the white part of $\mathcal{K}$ without intersecting γ_{t_1} must have diameter at least $t_1^{0.7}$. Of course all the choices above can be made $\mathcal{F}_{t_1}$-measurable.

Now it is $\bar{\gamma}$'s turn to do the rest within time $t_2 = t_3 + 1$. Choose a point in γ_{t_1} that has distance at least $t^{0.5}$ to the complement of $\mathcal{K}$ such that at least one third of its $2R$-neighbourhood is black and white respectively. Fix t_3 large enough that the probability of the event that $\tilde{\Phi}_{t_3}(\bar{\gamma})$ has distance to this point less or equal to R (and $\tilde{\Phi}_{t_3}(\bar{\gamma})$ is long) is at least ϵ uniformly in $\bar{\gamma}$. So with probability at least 0.5ϵ a point say x in $\tilde{\Phi}_{t_3}(\bar{\gamma})$ has an environment of diameter $3R$ at least one percent of which is white and black respectively. Choose another point in $\tilde{\Phi}_{t_3}(\bar{\gamma})$ say y with distance 1 of x such that the subcurve (denoted $\check{\gamma}$) of $\tilde{\Phi}_{t_3}(\bar{\gamma})$ linking x and y has diameter 1 and observe now that the lemma follows from Theorem 3.3.1 because we can choose t_1 large enough for $\check{\gamma}$ not to reach $\mathcal{K}^C$ within the remaining time 1 with sufficiently large probability. With $C_{19} := t_1 + t_3 + 1$ the proof is complete. $\square$

Lemma 4.4.6 (p_3).
There is $p_3 > 0$ such that $\mathbb{P}\left[\tau^R(\Gamma, tv) \leq T\right] \leq \frac{1}{p_3}\mathbb{P}\left[\tau^R(\Gamma, tv) \leq T; \mathrm{diam}(\Gamma_{T+C_{18}}) \geq 1\right]$.

Proof: This a direct consequence of the fact, that the diameter of long curves uniformly has a chance to grow to infinity without being smaller than 1 after time C_{18}. If necessary we increase C_{18} (without changing notation). $\square$

Proof of Theorem 4.4.3: First we use Lemma 4.4.4 to estimate

$$\mathbb{P}\left[\Gamma_{T+C_{18}} \cap \partial K_R(tv) \neq \emptyset\right] \geq \mathbb{P}\left[\tau^R(\Gamma, tv) \leq T\right] \cdot \mathbb{P}\left[\Gamma_{T+C_{18}} \cap \partial K_R(tv) \neq \emptyset \,\middle|\, \tau^R(\Gamma, tv) \leq T\right]$$
$$\geq \mathbb{P}\left[\tau^R(\Gamma, tv) \leq T; \mathrm{diam}(\Gamma_{T+C_{18}}) \geq 1\right]$$
$$\cdot \underbrace{\mathbb{P}\left[\mathrm{diam}(\Gamma_{T+C_{18}}) \geq 1; \Gamma_{T+C_{18}} \cap \partial K_R(tv) \neq \emptyset \,\middle|\, \tau^R(\Gamma, tv) \leq T\right]}_{\geq p_1}$$

which implies with Lemma 4.4.6 that

$$\mathbb{P}\left[\tau^R(\Gamma, tv) \leq T\right]$$
$$\leq \frac{1}{p_3}\mathbb{P}\left[\tau^R(\Gamma, tv) \leq T; \operatorname{diam}(\Gamma_{T+C_{18}}) \geq 1\right] \leq \frac{1}{p_1p_3}\mathbb{P}\left[\Gamma_{T+C_{18}} \cap \partial K_R(tv) \neq \emptyset\right]$$
$$\leq \frac{1}{p_1p_3}\left(\mathbb{P}\left[\Gamma_{T+C_{18}} \cap \partial K_R(tv) \neq \emptyset; \operatorname{diam}(\Gamma_{T+C_{18}}) \geq 1\right] + \mathbb{P}\left[\operatorname{diam}(\Gamma_{T+C_{18}}) < 1\right]\right).$$

Usage of Lemma 1.3.13 and symmetry yields now that the first term of the latter equals

$$\frac{1}{p_1p_3}\mathbb{P}\left[\Gamma \cap \partial K_R(tv)_{T+C_{18}} \neq \emptyset; \operatorname{diam}(\partial K_R(tv)_{T+C_{18}}) \geq 1\right]$$
$$\leq \frac{1}{p_1p_3}\mathbb{P}\left[\gamma \cap \partial K_R(tv)_{T+C_{18}+C_{19}} \neq \emptyset; \Gamma \cap \partial K_R(tv)_{T+C_{18}} \neq \emptyset\right]$$
$$/\mathbb{P}\left[\gamma \cap \partial K_R(tv)_{T+C_{18}+C_{19}} \neq \emptyset \mid \Gamma \cap \partial K_R(tv)_{T+C_{18}} \neq \emptyset; \operatorname{diam}(\partial K_R(tv)_{T+C_{18}}) \geq 1\right]$$
$$\leq \frac{1}{p_1p_3}\mathbb{P}\left[\gamma \cap \partial K_R(tv)_{T+C_{18}+C_{19}} \neq \emptyset\right]$$
$$/\mathbb{P}\left[\gamma \cap \partial K_R(tv)_{T+C_{18}+C_{19}} \neq \emptyset \mid \Gamma \cap \partial K_R(tv)_{T+C_{18}} \neq \emptyset; \operatorname{diam}(\partial K_R(tv)_{T+C_{18}}) \geq 1\right].$$

Applying Lemma 4.4.5 conditioned on $\mathcal{F}_{T+C_{18}}$ we obtain that this is less or equal than

$$\frac{1}{p_1p_2p_3}\mathbb{P}\left[\gamma \cap \partial K_R(tv)_{T+C_{18}+C_{19}} \neq \emptyset\right] = \frac{1}{p_1p_2p_3}\mathbb{P}\left[\gamma_{T+C_{18}+C_{19}} \cap \partial K_R(tv) \neq \emptyset\right]$$
$$\leq \frac{1}{p_1p_2p_3}\left(\mathbb{P}\left[\tau^R(\gamma, tv) \leq T + C_{18} + C_{19}\right] + \mathbb{P}\left[\operatorname{diam}(\gamma_{T+C_{18}+C_{19}}) < 1\right]\right)$$

where again we used Lemma 1.3.13. The fact the we are only considering $T \geq \sqrt{t}$ now shows that for $m \in \mathbb{N}$ there is $\kappa_m^{(11)} \in \mathbb{R}$ such that

$$\mathbb{P}\left[\operatorname{diam}(\gamma_{T+C_{18}+C_{19}}) < 1\right] \vee \mathbb{P}\left[\operatorname{diam}(\Gamma_{T+C_{18}}) < 1\right] \leq \kappa_m^{(11)}T^{-m}$$

which completes the proof (choosing $C_{20} := p_1p_2p_3$ and $C_{17} := C_{18} + C_{19}$). □

4.4 THE UPPER BOUND

If we consider now for $t \geq 2\frac{C_{17}}{\delta}$

$$\mathbb{P}\left[\tau^R(\Gamma, tv) \leq (1-\delta)t\right] \leq \frac{1}{C_{20}}\mathbb{P}\left[\tau^R(\gamma^{(t)}, tv) \leq (1-\delta)t + C_{17}\right] + \frac{\kappa_m^{(11)}t^{-m}}{C_{20}}$$

$$\leq \frac{1}{C_{20}}\underbrace{\mathbb{P}\left[\tau^R(\gamma^{(t)}, tv) \leq (1-\frac{\delta}{2})t\right]}_{\to 0 \text{ as } t \to \infty} + \frac{\kappa_m^{(11)}t^{-m}}{C_{20}} \to 0$$

together with $\mathbb{P}\left[\tau^R(\Gamma, tv) \leq (1+\delta)t\right] \geq \mathbb{P}\left[\tau^R(\gamma^{(t)}, tv) \leq (1+\delta)t\right] \to 1$ we get sending $t \to \infty$ that $\tau^R(\Gamma, tv)t^{-1}$ converges to 1 in probability. The diffeomorphic property of the flow of course implies that this convergence holds uniformly in $\gamma \in \mathcal{C}_R$ if we replace Γ by γ. Corollary 4.3.7 also shows that it also holds in L^p for any $p \geq 1$. We thus proved the following corollary.

Corollary 4.4.7 (convergence in probability).
We have for $\epsilon > 0$ uniformly in $\gamma \in \mathcal{C}_R$ that $\lim_{t\to\infty} \mathbb{P}\left[\left|\frac{\tau^R(\gamma,tv)}{t} - \|v\|^R\right| > \epsilon\right] = 0$
as well as for any $p > 0$ that $\lim_{t\to\infty} \mathbb{E}\left[\left|\frac{\tau^R(\gamma,tv)}{t} - \|v\|^R\right|^p\right] = 0.$

Proof: There is nothing left to show since we ensured that the assertions above do not depend on R. □

We proceed by proving the following version of the upper bound which directly implies 3. of Theorem 4.2.1.

Lemma 4.4.8 (convergence of probability).
We have for any $\epsilon > 0$ that $\lim_{t\to\infty} \mathbb{P}\left[\mathcal{W}_t^R(\gamma) \subset (1+\epsilon)t\mathcal{B}\right] = 1.$

Proof: We have equivalence with

$$\forall \epsilon > 0 : \lim_{t\to\infty} \mathbb{P}\left[\exists x \in \mathcal{W}_t^R(\gamma) : x \notin (1+\epsilon)t\mathcal{B}\right] = 0$$

$$\Leftarrow \forall \epsilon > 0 : \lim_{t\to\infty} \mathbb{P}\left[\exists x \in \mathbb{R}^2 : \tau^R(\gamma, x) \leq t; \frac{x}{t} \notin (1+\epsilon)\mathcal{B}\right] = 0$$

$$\Leftrightarrow \forall \epsilon > 0 : \lim_{t\to\infty} \mathbb{P}\left[\exists x \in \mathbb{R}^2 : \tau^R(\gamma, x) \leq t; \left\|\frac{x}{t(1+\epsilon)}\right\|^R > 1\right] = 0$$

$$\Leftrightarrow \forall \epsilon > 0 : \lim_{t\to\infty} \mathbb{P}\left[\exists x \in \mathbb{R}^2 : \tau^R(\gamma, (1+\epsilon)tx) \leq t; \|x\|^R > 1\right] = 0$$

$$\Leftrightarrow \forall \epsilon > 0 : \lim_{t\to\infty} \mathbb{P}\left[\exists x \in \mathbb{R}^2 : \tau^R(\gamma, tx) \leq \frac{t}{1+\epsilon}; \|x\|^R > 1\right] = 0.$$

So it is enough to show for $\epsilon > 0$ that $\lim_{t\to\infty} \mathbb{P}\left[\exists x \in \mathbb{R}^2 : \tau^R(\gamma, tx) \leq \frac{t}{1+\epsilon}; \|x\|^R = 1\right] = 0$. Choose $\delta \ll \epsilon$ and a δ-net on $\partial \mathcal{B}$ denoted by $\{v_j : j = 1, \ldots, N_\delta\}$. Then we can apply Corollary 4.4.7 to obtain $\lim_{t\to\infty} \mathbb{P}\left[\exists j \in \{1, \ldots, N_\delta\} : \frac{\tau^R(\gamma, tv_j)}{t} < \frac{1}{1+\frac{\epsilon}{2}}\right] =: \lim_{t\to\infty} \mathbb{P}\left[F_1(t)\right] = 0$. Due to Theorem 4.3.1 and Lemma 4.3.10 (similarly to the proof of Theorem 4.3.9) we have for large t (because for $v \in \mathbb{R}^2$ with $\|v\|^R = 1$ there is j with $\|v - v_j\| \leq \delta$) that

$$\mathbb{P}\left[F_1(t) \,|\, F_2(t)\right] := \mathbb{P}\left[\exists j \in \{1, \ldots, N_\delta\} : \frac{\tau^R(\gamma, tv_j)}{t} < \frac{1}{1+\frac{\epsilon}{2}} \,\Big|\, \exists v \in \partial \mathcal{B} : \frac{\tau^R(\gamma, tv)}{t} < \frac{1}{1+\epsilon}\right]$$

$$\geq \inf_{\gamma \in \mathcal{C}_{\frac{R}{2}} + tv, v \in \partial \mathcal{B}} \mathbb{P}\left[\tilde{\tau}^{\delta t}(\gamma, tv) < \left(\frac{1}{1+\frac{\epsilon}{2}} - \frac{1}{1+\epsilon}\right)t\right]$$

$$= \inf_{\gamma \in \mathcal{C}_{\frac{R}{2}}} \mathbb{P}\left[\tilde{\tau}^{\delta t}(\gamma, 0) < \left(\frac{1}{1+\frac{\epsilon}{2}} - \frac{1}{1+\epsilon}\right)t\right] \geq 1 - \kappa_m^{(12)} t^{-m} \qquad (4.27)$$

where $\kappa_m^{(12)} \in \mathbb{R}$ exists for $m \in \mathbb{N}$, provided $\delta = \delta(\epsilon)$ is chosen sufficiently small. This implies

$$\limsup_{t\to\infty} \mathbb{P}\left[F_2(t)\right] \leq \limsup_{t\to\infty}\left(\mathbb{P}\left[F_2(t)\,|\,F_1(t)\right] \underbrace{\mathbb{P}\left[F_1(t)\right]}_{\to 0} + \mathbb{P}\left[F_2(t); F_1^C(t)\right]\right)$$

$$\leq \limsup_{t\to\infty} \mathbb{P}\left[F_1^C(t) \,|\, F_2(t)\right] \mathbb{P}\left[F_2(t)\right] \leq \limsup_{t\to\infty} \kappa_m^{(12)} t^{-m} = 0 \qquad (4.28)$$

and hence Lemma 4.4.8. $\square$

Corollary 4.4.9 (2. of Theorem 4.2.1).
There is $\mathbb{P}$-a.s. for any $\epsilon > 0$ a sequence $(t_k : k \in \mathbb{N})$ with $t_k \nearrow \infty$ that fulfils

$$\mathcal{W}_{t_k}(\gamma) \subset (1+\epsilon) t_k \mathcal{B}.$$

Proof: With Lemma 4.4.8 this is a direct consequence of the fact, that convergence in probability implies a.s.-convergence of a subsequence. $\square$

Chapter 5

Dreaming Of Pesin's Formula

5.1 Introduction

Pesin's formula (see [43] and [44]) asserts that the entropy of a (random) dynamical system equals the sum of its positive Lyapunov characteristic numbers counted with their multiplicities. It has been extended to cover the case of dynamical systems preserving a Borel probability measure (see [15], [22] and [50]). The possible occurrence of zero characteristic exponents was finally captured in the papers [61], [31], [32], [33] and transferred to the random case in [34]. All these results as well as newer expositions of the topic e.g. [37] or [4] request the state space of the RDS to be a compact Riemannian manifold which of course guarantees the existence of an invariant probability (see e.g. [23, Theorem 1.4.3]). This existence is necessary to even define a notion of entropy so we focus on IOUFs here and postpone the discussion of the problem in the case of an IBF (or even in the case of an RIF). Since IOUFs possess nice smooth invariant probabilities we are left with the problem that the state space of an IOUF is $\mathbb{R}^d$ which is not compact at all. One straightforward approach is to throw the IOUF in $\mathbb{R}^d$ on the unit ball in $\mathbb{R}^{d+1}$ via a Lyapunov cohomology, apply the results given above and transfer them back to the original case. The aim of this chapter is to indicate why this does not work. Since its purpose is essentially the motivation of the treatment in Chapter 6 the reader may wish to omit it. We will not use it in the sequel.

5.2 Throwing An IOUF On A Ball - Stereographic Projection

We start with the definition of the Lyapunov cohomology we wish to consider.

Definition 5.2.1 (unit ball version of ϕ).
Let ψ be an IOUF with drift c and covariance tensor b. Denote by $K_r(x)$ the closed r-ball centered at x in $\mathbb{R}^{d+1}$ and define

$$g : \mathbb{R}^d \cup \{\infty\} \to K_1(0), \begin{cases} (x_1, \ldots, x_d) \mapsto (0, \ldots, 0, 1) + \frac{4}{4+|x|^2}(x_1, \ldots, x_d, -2) \\ \infty \mapsto (0, \ldots, 0, 1) \end{cases}$$

with the inverse

$$g^{-1} : K_1(0) \to \mathbb{R}^d \cup \{\infty\}, \begin{cases} (y_1, \ldots, y_{d+1}) \mapsto 2(\frac{y_1}{1-y_{d+1}}, \ldots, \frac{y_d}{1-y_{d+1}}) \\ (0, \ldots, 0, 1) \mapsto \infty \end{cases}$$

as well as $\Psi := g \circ (\mathbb{1}_{\mathbb{R}^d} \cdot \phi_t + \mathbb{1}_{\{\infty\}} \cdot \mathrm{id}_{\{\infty\}}) \circ g^{-1}$ which means that if we let

$$\Psi_t(x) = \begin{cases} g \circ \phi_t \circ g^{-1}(x) & : x \neq (0, \ldots, 0, 1) \\ (0, \ldots, 0, 1) & : x = (0, \ldots, 0, 1) \end{cases},$$

then $\Psi = \Psi_t$ is the unit ball version of ϕ.

We will focus on the case $t = 1$ because one can see all the important issues and it simplifies notation. Since we want to check whether Ψ is a diffeomorphism we have to consider an open domain U in $\mathbb{R}^d$ and a smooth diffeomorphism (map) f from U to an open neighbourhood of $(0, \ldots, 0, 1)$ and check if $\Xi := f^{-1} \circ \Psi \circ f$ is differentiable. We make the following choice.

$$f : \{x \in \mathbb{R}^d : |x| < 1\} \to \{y \in K_1(0) : y_{d+1} > 0\}, (x_1, \ldots, x_d) \mapsto (x_1, \ldots, x_d, \sqrt{1-|x|^2})$$

which yields

$$f^{-1} : \{y \in K_1(0) : y_{d+1} > 0\} \to \{x \in \mathbb{R}^d : |x| < 1\}, (y_1, \ldots, y_{d+1}) \mapsto (y_1, \ldots, y_d).$$

For we have $\Xi = f^{-1} \circ \Psi \circ f = f^{-1} \circ g \circ (\mathbb{1}_{\mathbb{R}^d} \cdot \phi_1 + \mathbb{1}_{\{\infty\}} \cdot \mathrm{id}_{\{\infty\}}) \circ g^{-1} \circ f$ we get $\Xi(0) = 0$

5.2 Throwing An IOUF On A Ball - Stereographic Projection

and for $x \neq 0$ that

$$\begin{aligned}
\Xi(x) &= f^{-1} \circ g \circ (\mathbb{1}_{\mathbb{R}^d} \cdot \phi_1 + \mathbb{1}_{\{\infty\}} \cdot \mathrm{id}_{\{\infty\}}) \circ g^{-1} \circ f(x) \\
&= f^{-1} \circ g \circ (\mathbb{1}_{\mathbb{R}^d} \cdot \phi_1 + \mathbb{1}_{\{\infty\}} \cdot \mathrm{id}_{\{\infty\}}) \circ g^{-1} \left(x_1, \ldots, x_d, \sqrt{1-|x|^2} \right) \\
&= f^{-1} \circ g \circ (\mathbb{1}_{\mathbb{R}^d} \cdot \phi_1 + \mathbb{1}_{\{\infty\}} \cdot \mathrm{id}_{\{\infty\}}) \left(\frac{2x_1}{1-\sqrt{1-|x|^2}}, \ldots, \frac{2x_d}{1-\sqrt{1-|x|^2}} \right) \\
&= f^{-1} \circ g \left(\phi_1 \left(\frac{2x}{1-\sqrt{1-|x|^2}} \right) \right) = f^{-1} \left((0,\ldots,0,1) + \frac{4\phi_1(\hat{x}(x))}{4+|\phi_1(\hat{x}(x))|^2}, -2 \right) \\
&= \frac{4}{4+|\phi_1(\hat{x}(x))|^2} \phi_1(\hat{x}(x))
\end{aligned}$$

wherein we put $\hat{x}(x) := g^{-1} \circ f(x)$. We fix some statements about $\hat{x}(x)$.

Proposition 5.2.2 (x and $\hat{x}(x)$).
$\hat{x}(x) = g^{-1} \circ f(x)$ satisfies the following.

1. $\hat{x}(x) = \frac{2x}{1-\sqrt{1-|x|^2}}$ and $x(\hat{x}) = \frac{4\hat{x}}{4+|\hat{x}|^2}$.

2. $\frac{\partial}{\partial x_i} \hat{x}_k(x) = \frac{2\delta_{ik}(1-\sqrt{1-|x|^2})\sqrt{1-|x|^2}-2x_ix_k}{(1-\sqrt{1-|x|^2})^2\sqrt{1-|x|^2}}$.

3. $|\hat{x}(x)| = \frac{2|x|}{1-\sqrt{1-|x|^2}}$.

4. $|x| \to 0 \Leftrightarrow |\hat{x}(x)| \to \infty$.

Proof: Since everything but 2. is clear, we only have to observe that

$$\frac{\partial}{\partial x_i} \hat{x}_k(x) = \frac{2\delta_{ik}(1-\sqrt{1-|x|^2}) + 2x_k \frac{\partial}{\partial x_i}\sqrt{1-|x|^2}}{(1-\sqrt{1-|x|^2})^2}$$

$$\frac{2\delta_{ik}(1-\sqrt{1-|x|^2}) + 2x_k \frac{-2x_i}{2\sqrt{1-|x|^2}}}{(1-\sqrt{1-|x|^2})^2} = \frac{2\delta_{ik}(1-\sqrt{1-|x|^2})\sqrt{1-|x|^2}-2x_ix_k}{(1-\sqrt{1-|x|^2})^2\sqrt{1-|x|^2}}.$$

□

We have to check Ξ for smoothness at 0.

Theorem 5.2.3 (smoothness problems).
The following holds.

1. Ξ is differentiable at 0 and we have $D_0\Xi = e^c \mathrm{id}_{\mathbb{R}^d}$.

2. The derivative of Ξ is not continuous at 0.

Proof: We start with the proof of 1. with the following observation for $x \in \mathbb{R}^d$.

$$\frac{\Xi(x) - \Xi(0) - e^c \mathrm{id}_{\mathbb{R}^d}(x)}{|x|} = \frac{1}{|x|}\left(\frac{4\phi_1(\hat{x}(x))}{4+|\phi_1(\hat{x}(x))|^2} - e^c x\right)$$

$$= \frac{1}{|x|}\left(\frac{4e^{-c}\hat{x}(x)}{4+e^{-2c}|\hat{x}(x)|^2} - e^c x + \frac{4\phi_1(\hat{x}(x))}{4+|\phi_1(\hat{x}(x))|^2} - \frac{4e^{-c}\hat{x}(x)}{4+e^{-2c}|\hat{x}(x)|^2}\right)$$

$$= \frac{1}{|x|}\frac{4e^{-c}\hat{x}(x) - (4+e^{-2c}|\hat{x}(x)|^2)e^c x}{4+e^{-2c}|\hat{x}(x)|^2} + \frac{1}{|x|}\left(\frac{4\phi_1(\hat{x}(x))}{4+|\phi_1(\hat{x}(x))|^2} - \frac{4e^{-c}\hat{x}(x)}{4+e^{-2c}|\hat{x}(x)|^2}\right) =: I + II$$

Since we have using Proposition 5.2.2 that

$$I = \frac{4+|\hat{x}(x)|^2}{4|\hat{x}(x)|}\frac{4e^{-c}\hat{x}(x) - (4+e^{-2c}|\hat{x}(x)|^2)e^c \frac{4\hat{x}(x)}{4+|\hat{x}(x)|^2}}{4+e^{-2c}|\hat{x}(x)|^2}$$

$$= \frac{4+|\hat{x}(x)|^2}{4|\hat{x}(x)|}\frac{4e^{-c}\hat{x}(x) - \frac{16e^c \hat{x}(x)}{4+|\hat{x}(x)|^2} - 4e^{-c}\frac{|\hat{x}(x)|^2}{4+|\hat{x}(x)|^2}\hat{x}(x)}{4+e^{-2c}|\hat{x}(x)|^2}$$

$$= \frac{4+|\hat{x}(x)|^2}{4|\hat{x}(x)|^2}|\hat{x}(x)|\frac{16e^{-c}\hat{x}(x) + 4e^{-c}|\hat{x}(x)|^2\hat{x}(x) - 16e^c \hat{x}(x) - 4e^{-c}|\hat{x}(x)|^2\hat{x}(x)}{(4+e^{-2c}|\hat{x}(x)|^2)(4+|\hat{x}(x)|^2)} \xrightarrow{|x|\to 0} 0$$

as well as

$$II = \frac{4+|\hat{x}(x)|^2}{4|\hat{x}(x)|}\frac{(16+4e^{-2c}|\hat{x}(x)|^2)\phi_1(\hat{x}(x)) - (16+4e^{-c}|\phi_1(\hat{x}(x))|^2)e^{-c}\hat{x}(x)}{(4+|\phi_1(\hat{x}(x))|^2)(4+e^{-2c}|\hat{x}(x)|^2)}$$

$$= \frac{4+|\hat{x}(x)|^2}{4+e^{-2c}|\hat{x}(x)|^2}\frac{16(\phi_1(\hat{x}(x)) - e^{-c}\hat{x}(x)) + 4(e^{-2c}|\hat{x}(x)|^2\phi_1(\hat{x}(x)) - e^{-c}|\phi_1(\hat{x}(x))|^2\hat{x}(x))}{4|\hat{x}(x)|(4+|\phi_1(\hat{x}(x))|^2)}$$

$$= \frac{4+|\hat{x}(x)|^2}{4+e^{-2c}|\hat{x}(x)|^2}\left(\frac{16(\phi_1(\hat{x}(x)) - e^{-c}\hat{x}(x))}{4|\hat{x}(x)|(4+|\phi_1(\hat{x}(x))|^2)} + e^{-2c}\frac{|\hat{x}(x)|(\phi_1(\hat{x}(x)) - e^{-c}\hat{x}(x))}{4+|\phi_1(\hat{x}(x))|^2}\right.$$

$$\left. + \frac{e^{-2c}|\hat{x}(x)|^2 - |\phi_1(\hat{x}(x))|^2}{4|\hat{x}(x)|(4+|\phi_1(\hat{x}(x))|^2)}e^{-c}\hat{x}(x)\right)$$

we get 1. by Lemma 2.2.1. To prove 2. we start by observing that $\Xi_j(x) = \frac{4\phi_1^j(\hat{x}(x))}{4+|\phi_1(\hat{x}(x))|^2}$

5.2 Throwing An IOUF On A Ball - Stereographic Projection

implies for $1 \leq i, j \leq d$ that

$$\frac{\partial}{\partial x_i} \Xi_j$$

$$= \frac{4}{4+|\phi_1(\hat{x}(x))|^2} \frac{\partial}{\partial x_i} \left[\phi_1^j(\hat{x}(x))\right] + \frac{\partial}{\partial x_i} \left[\frac{4}{4+|\phi_1(\hat{x}(x))|^2}\right] \phi_1^j(\hat{x}(x))$$

$$= \frac{4}{4+|\phi_1(\hat{x}(x))|^2} \frac{\partial}{\partial x_i} \left[\phi_1^j(\hat{x}(x))\right] - \frac{4\frac{\partial}{\partial x_i}[|\phi_1(\hat{x}(x))|^2]}{(4+|\phi_1(\hat{x}(x))|^2)^2} \phi_1^j(\hat{x}(x))$$

$$= \frac{4}{4+|\phi_1(\hat{x}(x))|^2} \left[\frac{\partial}{\partial x_i} \left[\phi_1^j(\hat{x}(x))\right] - \frac{2\phi_1^j(\hat{x}(x))}{4+|\phi(\hat{x}(x))|^2} \sum_k \phi_1^k(\hat{x}(x)) \frac{\partial}{\partial x_i}(\phi_1^k(\hat{x}(x)))\right]$$

$$= \frac{4\left[\sum_l \partial_l \phi_1^j(\hat{x}(x)) \frac{\partial}{\partial x_i}\hat{x}_l(x) - \frac{2\phi_1^j(\hat{x}(x))}{4+|\phi_1(\hat{x}(x))|^2} \sum_k \phi_1^k(\hat{x}(x)) \sum_l \partial_l \phi_1^k(\hat{x}(x)) \frac{\partial}{\partial x_i}\hat{x}_l(x)\right]}{4+|\phi_1(\hat{x}(x))|^2}$$

$$= \frac{4}{4+|\phi_1(\hat{x}(x))|^2} \sum_l \frac{\partial}{\partial x_i} \hat{x}_l(x) \left[\partial_l \phi_1^j(\hat{x}(x)) - 2\phi_1^j(\hat{x}(x)) \sum_k \frac{\phi_1^k(\hat{x}(x))\partial_l \phi_1^k(\hat{x}(x))}{4+|\phi_1(\hat{x}(x))|^2}\right].$$

By Proposition 5.2.2 this equals

$$\frac{4}{4+|\phi_1|^2} \sum_l \frac{2\delta_{il}(1-\sqrt{1-|x|^2})\sqrt{1-|x|^2} - 2x_i x_l}{(1-\sqrt{1-|x|^2})^2\sqrt{1-|x|^2}} \left[\partial_l \phi_1^j - 2\phi^j \sum_k \frac{\phi_1^k \partial_l \phi_1^k}{4+|\phi_1|^2}\right]$$

wherein we write ϕ_1^j for $\phi_1^j(\hat{x}(x))$, $\partial_l \phi_1^j$ for $\partial_l \phi_1^j(\hat{x}(x))$ etc.. Since we obviously have the Taylor-expansions (as $r \to 0$)

$$\sqrt{1-r^2} \approx 1 - \frac{1}{2}r^2 - \frac{1}{8}r^4,$$

$$1 - \sqrt{1-r^2} \approx \frac{1}{2}r^2 + \frac{1}{8}r^4,$$

$$(1-\sqrt{1-r^2})\sqrt{1-r^2} \approx \frac{1}{2}r^2 - \frac{1}{8}r^4,$$

$$(1-\sqrt{1-r^2})^2\sqrt{1-r^2} \approx \frac{1}{4}r^4 - \frac{1}{64}r^8,$$

$$(1-\sqrt{1-r^2})^2 \approx \frac{1}{4}r^4 + \frac{1}{8}r^6$$

we get as $x \to 0$ for the special choice $x = (r, 0, \ldots, 0)$ and $i = 1$ that

$$\frac{2\delta_{il}(1-\sqrt{1-|x|^2})\sqrt{1-|x|^2} - 2x_i x_l}{(1-\sqrt{1-|x|^2})^2\sqrt{1-|x|^2}} = \frac{2\delta_{1l}(1-\sqrt{1-r^2})\sqrt{1-r^2} - 2r^2\delta_{1l}}{(1-\sqrt{1-r^2})^2\sqrt{1-r^2}}$$

$$\approx \delta_{1l} \frac{-\frac{1}{4}r^4 - r^2}{\frac{1}{4}r^4 - \frac{1}{64}r^8} \approx -4\delta_{1l}r^{-2}$$

$$\frac{4}{4+|\phi_1|^2} \sim \frac{4}{4+e^{-2c}|\hat{x}(x)|^2} = \frac{4}{4+e^{-2c}|\frac{2r}{1-\sqrt{1-r^2}}|^2}$$

$$\approx \frac{4}{4+e^{-2c}\frac{4r^2}{\frac{1}{4}r^4+\frac{1}{8}r^6}} \approx \frac{1}{4}e^{2c}r^2$$

wherein we used Lemma 2.2.1 and Proposition 5.2.2. This altogether yields

$$\frac{\partial}{\partial x_1}\Xi_j \approx -e^{2c}\left[\partial_1 \phi_1^j - 2\phi_1^j \sum_k \frac{\phi_1^k \partial_1 \phi_1^k}{4+|\phi_1|^2}\right]$$

which means that it is sufficient to show that $\partial_1 \phi_1^2 - \phi_1^2 \sum_k \frac{\phi_1^k \partial_1 \phi_1^k}{4+|\phi_1|^2}$ does not converge to 0 w.p.p. as $r \to 0$. Since we have $\frac{\phi_1^2 \phi_1^k}{4+|\phi_1|^2} \sim \delta_{k1}\delta_{21} = 0$ we thus have to show that $\partial_1 \phi_1^2$ does not converge to 0 w.p.p as $x \to 0$. Since we assumed (1.6) we get that spatial derivatives of ϕ become approximately independent if taken in points far from each other and thus we see that $\partial_1 \phi_1^2 = \partial_1 \phi_1^2(\hat{x}(x)) = \partial_1 \phi_1^2(\frac{(r,0,\ldots,0)}{1-\sqrt{1-r^2}})$ does not converge with probability 1 because $\lim_{r \to 0}\left|\frac{(r,0,\ldots,0)}{1-\sqrt{1-r^2}}\right| = \infty$. $\square$

Hence we will have to work on our own to obtain results like Pesin's formula in the situation of an IOUF.

Chapter 6

The Margulis-Ruelle Inequality for IOUFs

As we saw in Chapter 5 we will have to prove the results indicated there in the case of an IOUF on our own. We start with the $\leq$-part in Pesin's formula.

6.1 Introduction

Ruelle's inequality (sometimes also called Margulis-Ruelle inequality) asserts that the entropy of a (random) dynamical system can be bounded from above by the sum of its positive Lyapunov characteristic numbers counted with their multiplicities. It has been known for measure-preserving C^1-maps since the late 1970's (see [49]) and was generalized to capture certain singularities in [22]. It was first formulated in the case of a sequence of i.i.d. random transformations in [24]. The fact that this proof contained a gap lead to the two independent corrections namely [37] and [3]. The latter generalized the inequality to the case of a RDS on a compact Riemannian manifold. This compactness assumption makes the standard literature cited above not directly applicable to the case of an IOUF because its state space $\mathbb{R}^d$ is not compact and it also fails to satisfy assumptions on boundedness or uniform continuity. As a compactification via stereographic projection fails to yield the C^1-property, we need to work a bit to get the inequality. It turns out that most of the ideas of [24] and of [3] can be transferred to the case of an IOUF. This shows that Ruelle's inequality can hold even in the non-compact case.

6.2 Basics From Entropy Theory

This section deals with basic entropy theory and its application to the case of an i.i.d. sequence of random diffeomorphisms obtained from an IOUF. Although the concepts are quite elementary we will go through the entire construction for completeness. Note that the state space $\mathbb{R}^d$ is not compact so special care has to be taken concerning existence of invariant probabilities because the classical Krylov-Bogoljubov result does not apply (cf. [58, Theorem 5.13] or [24, Lemma 2.2.2]). This is exactly what we did in Proposition 1.4.5. In this whole chapter we will consider a sequence $\phi_{0,1}, \phi_{1,2}, \ldots$ of i.i.d. smooth random diffeomorphisms with law $m := \mathcal{L}(\phi_{0,1}(\cdot))$ where ϕ is an IOUF as constructed in Definition 1.4.1. W.l.o.g. we will assume that the $\phi_{i,i+1}$'s are defined on $(\Omega, \mathcal{F}, \mathbb{P}) := ((\mathrm{Diff}(\mathbb{R}^d))^{\mathbb{N}}, \mathcal{B}(\mathrm{Diff}(\mathbb{R}^d))^{\mathbb{N}}, m^{\otimes \mathbb{N}})$ by $\phi_{n-1,n}(\omega) = \omega(n)$. We will also fix the shift $\theta \colon \Omega \to \Omega$ as

$$\theta \colon \Omega \to \Omega, (\phi_{0,1}, \phi_{1,2}, \phi_{2,3}, \ldots) \mapsto (\phi_{1,2}, \phi_{2,3}, \ldots) \text{ i.e. } \theta(\omega)_n = \omega_{n+1}$$

as well as the skew-product shift $\tau \colon \Omega \times \mathbb{R}^d \to \Omega \times \mathbb{R}^d$ as

$$\tau(\omega, z) := (\theta\omega, \omega_1(z)).$$

On $\mathrm{Diff}(\mathbb{R}^d)$ we may choose any topology such that the mapping $(f, x) \mapsto f(x)$ is $\mathcal{B}(\mathrm{Diff}(\mathbb{R}^d)) \otimes \mathcal{B}(\mathbb{R}^d) - \mathcal{B}(\mathbb{R}^d)$ measurable. For instance we may take a localized C^1-topology. Of course the random variables $\phi_{0,1}, \phi_{1,2}, \ldots$ form an i.i.d. sequence with $\phi_{i-1,i}(\theta\omega) = \phi_{i,i+1}(\omega)$ i.e. we have $\phi_{i-1,i} \circ \theta = \phi_{i,i+1}$. The concepts we introduce in the sequel are widely known, we follow [24] in the exposition.

6.2.1 The Definition Of The Metric Entropy Of An IOUF

We start with the definition of the entropy of a partition.

Definition 6.2.1 (entropy of a partition).
Let ξ be a countable partition of the probability space $(\Omega, \mathcal{F}, \mathbb{P})$. Let $\mathcal{G} \subset \mathcal{F}$ be a sub-σ-algebra of $\mathcal{F}$. The conditional entropy of ξ given $\mathcal{G}$ is the number

$$H_{\mathbb{P}}(\xi \,|\, \mathcal{G}) := -\int_{\Omega} \sum_{A \in \xi} \mathbb{P}[A \,|\, \mathcal{G}] \log \mathbb{P}[A \,|\, \mathcal{G}] \, d\mathbb{P}.$$

6.2 Basics From Entropy Theory

The number

$$H_\mathbb{P}\left(\xi\,|\{\Omega,\emptyset\}\right) := -\sum_{A\in\xi} \mathbb{P}\left[A\right]\log \mathbb{P}\left[A\right] \in [0,\infty]$$

is called the entropy of ξ.

We let $0\log 0 = 0$ and so the sums in the previous definition always make sense. Note that the function $\iota : x \mapsto x\log x$ is strictly convex on $[0,\infty)$. We now state some basic properties of the entropy. For two partitions ξ and η we denote their common refinement by $\xi \vee \eta$. Note also that $\sigma(\xi)$ denotes the σ-algebra generated by the elements of ξ.

Lemma 6.2.2 (trivial entropy bound).
The entropy of a partition is at most the logarithm of its cardinality.

Proof: The infinite case is trivial and the finite case is proved to be an easy consequence of Jensen's inequality in [24, Corollary 2.1.1]. □

Lemma 6.2.3 (properties of the entropy of a partition).
Let ξ and η be countable partitions of the probability space $(\Omega, \mathcal{F}, \mathbb{P})$. Let further $\tilde{\mathcal{G}}, \mathcal{G} \subset \mathcal{F}$ be σ-algebras and $f : (\Omega, \mathcal{F}, \mathbb{P}) \to (\Omega, \mathcal{F}, \mathbb{P})$ be an endomorphism i.e. a deterministic measure preserving measurable map. Then the following holds true.

1. $H_\mathbb{P}\left(\xi\,|\mathcal{G}\right) \geq 0$.
2. $H_\mathbb{P}\left(\xi \vee \eta\,|\mathcal{G}\right) = H_\mathbb{P}\left(\xi\,|\mathcal{G}\right) + H_\mathbb{P}\left(\eta\,|\sigma(\xi) \vee \mathcal{G}\right)$.
3. $H_\mathbb{P}\left(\xi \vee \eta\right) = H_\mathbb{P}\left(\xi\right) + H_\mathbb{P}\left(\eta\,|\sigma(\xi)\right)$.
4. $\xi \prec \eta \Rightarrow H_\mathbb{P}\left(\xi\,|\mathcal{G}\right) \leq H_\mathbb{P}\left(\eta\,|\mathcal{G}\right)$.
5. $\xi \prec \eta \Rightarrow H_\mathbb{P}\left(\xi\right) \leq H_\mathbb{P}\left(\eta\right)$.
6. $\tilde{\mathcal{G}} \subset \mathcal{G} \Rightarrow H_\mathbb{P}\left(\xi\,|\mathcal{G}\right) \leq H_\mathbb{P}\left(\eta\,|\tilde{\mathcal{G}}\right)$.
7. $H_\mathbb{P}\left(\xi\right) \geq H_\mathbb{P}\left(\xi\,|\mathcal{G}\right)$.
8. $H_\mathbb{P}\left(\xi \vee \eta\,|\mathcal{G}\right) \leq H_\mathbb{P}\left(\xi\,|\mathcal{G}\right) + H_\mathbb{P}\left(\eta\,|\mathcal{G}\right)$.
9. $H_\mathbb{P}\left(\xi \vee \eta\right) \leq H_\mathbb{P}\left(\xi\right) + H_\mathbb{P}\left(\eta\right)$.

10. $H_\mathbb{P}\left(f^{-1}\xi \mid f^{-1}\mathcal{G}\right) = H_\mathbb{P}\left(\xi \mid \mathcal{G}\right)$.

11. $H_\mathbb{P}\left(f^{-1}\xi\right) = H_\mathbb{P}\left(\xi\right)$.

Proof: [24, Remark 2.1.1 and Lemma 2.1.2]. Note that the fact that the results are stated only for finite partitions there does not alter the proof at all. □

Definition and Lemma 6.2.4 (standard definition of entropy).
Let $f : (\Omega, \mathcal{F}, \mathbb{P}) \to (\Omega, \mathcal{F}, \mathbb{P})$ be an endomorphism and $\mathcal{G} \subset \mathcal{F}$ be a σ-algebra such that $f^{-1}\mathcal{G} \subset \mathcal{G}$. Then for any countable partition ξ there exists the limit

$$h_\mathbb{P}^\mathcal{G}(\xi, f) := \lim_{n \to \infty} \frac{1}{n} H_\mathbb{P}\left(\bigvee_{i=0}^{n-1} f^{-i}\xi \,\middle|\, \mathcal{G}\right).$$

This limit is called the entropy of f with respect to ξ given $\mathcal{G}$. The limit is finite provided one of the entries of the sequence is finite. Furthermore the numbers

$$h_\mathbb{P}^\mathcal{G}(f) := \sup_\xi h_\mathbb{P}^\mathcal{G}(\xi, f) \text{ and } h_\mathbb{P}(f) := \sup_\xi h_\mathbb{P}^{\{\emptyset, \Omega\}}(\xi, f)$$

are called the entropy of f given $\mathcal{G}$ and the entropy of f respectively. The suprema can be taken over all finite partitions, over all countable partitions with finite entropy or over all countable partitions (all choices lead to the same result).

Proof: [24, Theorem 2.1.1] and [48, Paragraph 9]. □

We state some properties of the entropy that do not depend on compactness assumptions on the state space.

Lemma 6.2.5 (properties of the entropy of an endomorphism).
Suppose that ξ and η are countable partitions of Ω, that $f : \Omega \to \Omega$ is an endomorphism and that $\mathcal{G} \subset \mathcal{F}$ is a σ-algebra such that $f^{-1}\mathcal{G} \subset \mathcal{G}$. Then the following holds.

1. $h_\mathbb{P}^\mathcal{G}(f, \xi) \leq H_\mathbb{P}\left(\xi \mid \mathcal{G}\right)$.

2. $h_\mathbb{P}^\mathcal{G}(f, \xi \vee \eta) \leq h_\mathbb{P}^\mathcal{G}(f, \xi) + h_\mathbb{P}^\mathcal{G}(f, \eta)$.

3. $\xi \prec \eta \Rightarrow h_\mathbb{P}^\mathcal{G}(f, \xi) \leq h_\mathbb{P}^\mathcal{G}(f, \eta)$.

6.2 Basics From Entropy Theory

4. $h_\mathbb{P}^\mathcal{G}(f,\xi) \leq h_\mathbb{P}^\mathcal{G}(f,\xi) + H_\mathbb{P}(\xi\,|\sigma(\eta) \vee \mathcal{G}) \leq h_\mathbb{P}^\mathcal{G}(f,\xi) + H_\mathbb{P}(\xi\,|\sigma(\eta))$.

5. $h_\mathbb{P}^\mathcal{G}(f, f^{-1}\xi) \leq h_\mathbb{P}^\mathcal{G}(f,\xi)$.

6. For $n \in \mathbb{N}$ we have $h_\mathbb{P}^\mathcal{G}(f,\xi) = h_\mathbb{P}^\mathcal{G}\left(f, \bigvee_{i=1}^{n-1} f^{-i}\xi\right)$.

7. For $n \in \mathbb{N}$ we have $h_\mathbb{P}^\mathcal{G}(f^n,\xi) = n h_\mathbb{P}^\mathcal{G}(f,\xi)$.

8. If ξ is an $\mathcal{G}$-generator i.e. if $\sigma\left(\mathcal{G} \vee \bigvee_{i=0}^\infty f^{-i}\xi\right) = \mathcal{F}$ mod $\mathbb{P}$ then $h_\mathbb{P}^\mathcal{G}(f) = h_\mathbb{P}^\mathcal{G}(f,\xi)$.

9. If $\xi_1 \prec \xi_2 \prec \ldots$ is a sequence of partitions such that $\mathcal{F} = \sigma(\mathcal{G} \vee \bigvee_{i=1}^\infty \xi_1)$ mod $\mathbb{P}$ then

$$h_\mathbb{P}^\mathcal{G}(f) = \lim_{n\to\infty} h_\mathbb{P}^\mathcal{G}(f,\xi_n).$$

10. $h_\mathbb{P}^\mathcal{G}(f,\xi) \leq H_\mathbb{P}(f^{-1}\xi\,|\sigma(\xi) \vee \mathcal{G})$.

Proof: [24, Lemmas 2.1.3, 2.1.4, 2.1.5 and 2.1.6] as well as [3, Lemma 2]. □

Up to now we did not specify how we want to link the concept of entropy of a measure preserving transformation to an IOUF. The above yields three straightforward (but frequently useless) choices. We will state them nevertheless because one of them can be rearranged to yield some meaningful concept.

1. One could consider the space $\mathbb{R}^d \times \Omega$ with the skew-product transformation τ from Definition and Lemma 1.2.4 (in the discrete time version) and the trivial σ-algebra.

2. We might also think of the state space Ω with the shift θ_1 as measure-preserving transformation (cf. (1.3)).

3. The third options is to take $(\mathbb{R}^d)^\mathbb{N}$ and to consider the shift on the set of possible outcomes of the unit step discretized one-point motion with the associated Markov measure.

Under certain assumptions (which are not satisfied for IOUFs and correspond to a certain fast delocalization) [24, Theorem 2.1.2] says that all the choices above yield a notion of entropy that equals infinity. Though the assumptions of this theorem are not satisfied we strongly conjecture that its assertion remains true and hence we need a more sophisticated notion of entropy.

Definition and Lemma 6.2.6 (metric entropy).
Let μ be an invariant probability for an IOUF (in the sense of Markov processes) and ξ be a countable partition of $\mathbb{R}^d$. Then there exists

$$h_\mu(\phi,\xi) := \lim_{n\to\infty} \frac{1}{n} \int H_\mu \left(\bigvee_{i=0}^{n-1} \phi_{0,i}(\cdot,\omega)^{-1}\xi \right) d\mathbb{P}(\omega).$$

The number

$$h_\mu(\phi) := \sup_\xi h_\mu(\phi,\xi)$$

is called the metric entropy of ϕ. The supremum is taken over all finite partitions, all countable ones or equivalently over all measurable partitions (see [48] for the definition of measurable partition).

Proof: [24, Theorem 2.1.3] and [48, Paragraph 9]. □

One might ask why we introduce various types of entropy, when conjecturing them to equal infinity. We now come to the answer to this question. The following theorem states that the metric entropy of an IOUF can be viewed as a special case of one of the entropy concepts given above (with a slight modification).

Theorem 6.2.7 (different versions of metric entropy).
Let μ be an invariant measure (in the sense of Markov processes) for the one-point motion of an IOUF ϕ. Let further $\mathcal{B}_\Omega$ be the σ-algebra generated by elements of the form $\{\omega \in \Omega : \phi_{0,1}(\omega) \in A_1, \ldots \phi_{n-1,n}(\omega) \in A_n\}$ for $n \in \mathbb{N}$ and $A_1, \ldots, A_n \in \mathcal{B}(\mathrm{Diff}(\mathbb{R}^d))$. Let $\mathbb{R}^d \times \mathcal{B}_\Omega := \{\mathbb{R}^d \times A : A \in \mathcal{B}_\Omega\}$. Consider the skew product transform $\tau : \Omega \times \mathbb{R}^d \to \Omega \times \mathbb{R}^d$ from Definition and Lemma 1.2.4. Then the following holds.

1. *If $\xi = \{A_1, \ldots, A_n\}$ and $\eta = \{B_1, \ldots, B_m\}$ are finite measurable partitions of $\mathbb{R}^d$ and Ω respectively, then*

$$h_\mu(\phi,\xi) = h_{\mu\otimes\mathbb{P}}^{\mathbb{R}^d \times \mathcal{B}_\Omega}(\tau, \xi \times \eta)$$

   where $\xi \times \eta := \{A_i \times B_j : 1 \leq i \leq n, 1 \leq j \leq m\}$.

2. $h_\mu(\phi) = h_{\mu\otimes\mathbb{P}}^{\mathbb{R}^d \times \mathcal{B}_\Omega}(\tau)$.

Proof: [24, Theorem 2.1.4]. □

The short form of the above is: the entropy of an IOUF can be written as the conditional entropy of the skew-product given the randomness.

Corollary 6.2.8 (properties of metric entropy).
Let μ be the invariant measure for the one-point motion of ϕ. Then we have the following.

1. For any $n \in \mathbb{N}$ we have
$$h_\mu(\phi_{0,n}) = n h_\mu(\phi_{0,1}).$$

2. If ξ is a finite partition of $\mathbb{R}^d$ such that for $\mathbb{P}$-a.e. ω $\sigma\left(\bigvee_{i=0}^\infty \phi_{0,i}(\cdot,\omega)^{-1}\xi\right)$ coincides mod μ with $\mathcal{B}(\mathbb{R}^d)$ then
$$h_\mu(\phi) = h_\mu(\phi, \xi).$$

3. If $\xi_1, \xi_2, \ldots$ is a sequence of countable partitions of finite entropy such that $\lim_{n\to\infty} \operatorname{diam}(\xi_n) = 0$ wherein $\operatorname{diam}(\xi_n) = \sup_{C \in \xi_n} \operatorname{diam}(C)$ then
$$h_\mu(\phi) = \lim_{n\to\infty} h_\mu(\phi, \xi_n).$$

Proof: This follows from Lemma 6.2.5, [37, Theorem 0.4.7] and Theorem 6.2.7. □

Remark: Since we also have for any $t \geq 0$ that $h_\mu(\phi_{0,t}) = t h_\mu(\phi_{0,1})$ we see that discretizing the flow in unit steps is no restriction at all.

6.3 The Margulis-Ruelle Inequality

6.3.1 Statement Of The Result

In this section we want to prove one half of Pesin's formula i.e. the following theorem.

Theorem 6.3.1 (Margulis-Ruelle inequality).
Let ϕ be an IOUF with Lyapunov exponents $\lambda_1, \ldots, \lambda_d$. Then we have
$$h_\mu(\phi) \leq \sum_{i=1}^d \lambda_i^+ = \sum_{i=1}^d (\lambda_i \vee 0).$$

First we state a corollary to Lemma 6.2.5 which is taken from [3] where it is proven for the general case of RDS'.

Corollary 6.3.2 (entropy estimate via special countable partitions).
Let ξ be a countable partition of $\mathbb{R}^d$ and μ be as above. Then we have

$$h_\mu(\phi, \xi) \leq \int_\Omega H_\mu\left(\phi_{0,1}^{-1}(\cdot, \omega)\xi \big| \xi\right) d\mathbb{P}(\omega).$$

Proof: Since in the case of a product measure the disintegration is (in fact can be chosen to be) trivial we can simply apply [3, Corollary 1]. □

We shall also need the following purely deterministic result from geometry. Therefore we let $K_\epsilon(X) := \{y \in \mathbb{R}^d : \text{dist}(y, X) \leq \epsilon\}$ denote the closed ϵ-neighbourhood of $X \subset \mathbb{R}^d$.

Lemma 6.3.3 (geometric estimate).
Let $A : \mathbb{R}^d \to \mathbb{R}^d$ be a linear mapping and let $\mathbb{R}^d$ be equipped with the usual Euclidean norm $|\cdot|$. Let further $\delta_1(A) \geq \ldots \geq \delta_d(A)$ denote the singular values of A. Then there exists a constant $C(d)$ which only depends on d such that for any $\epsilon > 0$ the number of disjoint balls with radius $\frac{\epsilon}{2}$, which can intersect $K_{2\epsilon}(AK_\epsilon(0))$ does not exceed

$$C(d) \prod_{u=1}^d (\delta_u(A) \vee 1).$$

Proof: [30, Lemma II.2.3]. □

Now we can turn to the proof of Theorem 6.3.1 which will be carried out in the following two subsections.

6.3.2 Construction Of The Partitions

Fix $n \in \mathbb{N}$. For $k, l \in \mathbb{N}$ we define $\Omega_{k,l}$ to be the set of $\omega \in \Omega$ on which we have the following. For any $\epsilon < \frac{1}{k}$ and $x, y \in K_l(0)$ the inequality $|x - y| < \epsilon$ implies

$$|\phi_n(x) - \phi_n(y) - D_x\phi_n(\cdot)(y - x)| \leq \epsilon$$
$$\text{and for } i \in \{1, \ldots, d\} \text{ that } \frac{1}{2} \leq \frac{\delta_i(D_x\phi(\cdot)\vee)1}{\delta_i(D_y\phi(\cdot)\vee)1} \leq 2. \tag{6.1}$$

For $k, l \in \mathbb{N}$ we have the obvious inclusions $\Omega_{k,l+1} \subset \Omega_{k,l} \subset \Omega_{k+1,l}$ and for $l \in \mathbb{N}$ that $\Omega = \cup_{k=1}^\infty \Omega_{k,l}$. We choose a maximal $\frac{1}{k}$-separated set $E_{k,l} = \{x_1, \ldots, x_m\}$ in $K_l(0)$ (wherein of course m depends on k and l) and extend it to a maximal $\frac{1}{k}$-separated set $\{x_1, x_2, \ldots\}$ in $\mathbb{R}^d$. Herein a set is said to be ϵ-separated if no two of its elements have distance less then ϵ

6.3 THE MARGULIS-RUELLE INEQUALITY

and an ϵ-separated set is said to be maximal if any proper superset fails to be ϵ-separated. Letting further $\eta_i := \{y \in \mathbb{R}^d : \forall i \neq j \in \mathbb{N} : |y - x_i| < |y - x_j|\}$ we can finally define the partitions $\xi_k = \{\xi_{x_i} : i \in \mathbb{N}\}$ of $\mathbb{R}^d$ in the following way

$$\xi_{x_1} = \overline{\eta_1}, \xi_{x_i+1} = \overline{\eta_{i+1}} \setminus \bigcup_{j=1}^{i} \eta_j \text{ for } i \geq 2.$$

The dependence of ξ_k on k is hidden in the definitions of the x_i. We have for arbitrary $i \in \mathbb{N}$ that $\xi_{x_i} \subset K_{\frac{1}{k}}(x_i)$ which means that $\text{diam}(\xi_k) \leq \frac{1}{k}$.

6.3.3 Entropy Estimates

For $\xi_1, \xi_2, \ldots$ is a suitable sequence of countable measurable partitions, we can conclude with Corollary 6.2.8 and Corollary 6.3.2 that

$$nh_\mu(\phi) = h_\mu(\phi_n) = \lim_{k \to \infty} h_\mu(\phi_n, \xi_k) \leq \lim_{k \to \infty} \int_\Omega H_\mu\left(\phi_n^{-1}\xi_k \mid \xi_k\right) d\mathbb{P}.$$

For we have that

$$H_\mu\left(\phi_n^{-1}\xi_k \mid \xi_k\right) = -\sum_{i=1}^{\infty} \mu(\xi_{x_i}) \sum_{j=1}^{\infty} \mu(\phi_n^{-1}\xi_{x_j} \mid \xi_{x_i}) \log \mu(\phi_n^{-1}\xi_{x_j} \mid \xi_{x_i})$$

we can estimate for arbitrary $l = l(n)$

$$nh_\mu(\phi) \leq \lim_{k \to \infty} \int_\Omega H_\mu\left(\phi_n^{-1}\xi_k \mid \xi_k\right) d\mathbb{P}$$
$$= -\lim_{k \to \infty} \int_{\Omega_{k,l}} \sum_{i=1}^{m} \mu(\xi_{x_i}) \sum_{j=1}^{\infty} \mu(\phi_n^{-1}\xi_{x_j} \mid \xi_{x_i}) \log \mu(\phi_n^{-1}\xi_{x_j} \mid \xi_{x_i}) d\mathbb{P}$$
$$- \lim_{k \to \infty} \int_\Omega \sum_{i=m+1}^{\infty} \mu(\xi_{x_i}) \sum_{j=1}^{\infty} \mu(\phi_n^{-1}\xi_{x_j} \mid \xi_{x_i}) \log \mu(\phi_n^{-1}\xi_{x_j} \mid \xi_{x_i}) d\mathbb{P}$$
$$- \lim_{k \to \infty} \int_{\Omega \setminus \Omega_{k,l}} \sum_{i=1}^{m} \mu(\xi_{x_i}) \sum_{j=1}^{\infty} \mu(\phi_n^{-1}\xi_{x_j} \mid \xi_{x_i}) \log \mu(\phi_n^{-1}\xi_{x_j} \mid \xi_{x_i}) d\mathbb{P}$$
$$=: I + II + III.$$

(The following computations will show that the limits on the right hand side of the above exist separately.) We will estimate the number of elements of ξ_k that can intersect $\phi_n \xi_{x_i}$

to estimate I via Lemma 6.2.2. Since for $i \leq m$ and $\omega \in \Omega_{k,l}$ by (6.1) we have that

$$\phi_n \xi_i \subset \phi_n K_{\frac{1}{k}}(x_i) \subset K_{\frac{1}{k}}\left(\phi_n(x_i) + D_{x_i}\phi_n K_{\frac{1}{k}}(0)\right) = \phi_n(x_i) + K_{\frac{1}{k}}\left(D_{x_i}\phi_n K_{\frac{1}{k}}(0)\right)$$

we have for $i \leq m$, $\omega \in \Omega_{k,l}$ and $j \in \mathbb{N}$ that $\xi_{x_j} \cap \phi_n \xi_{x_i} \neq \emptyset$ implies

$$K_{\frac{1}{2k}}(x_j) \cap \phi_n(x_i) + K_{\frac{2}{k}}\left(D_{x_i}\phi_n K_{\frac{1}{k}}(0)\right) \neq \emptyset.$$

Therefore we have for $i \in \{1, \ldots, m\}$ and $\omega \in \Omega_{k,l}$ by Lemma 6.3.3 applied to $A := D_x \phi_n$

$$\#\{j : \xi_{x_j} \cap \phi_n(\cdot, \omega)\xi_{x_i} \neq \emptyset\} \leq K(n, \omega, x_i) := C(d) \prod_{u=1}^{d} (\delta_u(D_{x_i}\phi_n(\cdot, \omega)) \vee 1)$$

which implies by Lemma 6.2.2

$$I = -\lim_{k \to \infty} \int_{\Omega_{k,l}} \sum_{i=1}^{m} \mu(\xi_{x_i}) \sum_{j=1}^{\infty} \mu(\phi_n^{-1}\xi_{x_j}|\xi_{x_i}) \log \mu(\phi_n^{-1}\xi_{x_j}|\xi_{x_i}) d\mathbb{P}$$

$$\leq \lim_{k \to \infty} \int_{\Omega_{k,l}} \sum_{i=1}^{m} \mu(\xi_{x_i}) \log K(n, \omega, x_i) d\mathbb{P}(\omega)$$

$$= \lim_{k \to \infty} \int_{\Omega_{k,l}} \sum_{i=1}^{m} \int_{\xi_{x_i}} \log K(n, \omega, x_i) d\mu(y) d\mathbb{P}(\omega).$$

But since for $i \in \{1, \ldots, m\}$, $y \in \xi_{x_i}$ and $\omega \in \Omega_{k,l}$ we have

$$\log K(n, \omega, x_i) = \log C(d) + \sum_{u=1}^{d} \log^+(\delta_u(D_x \phi_n))$$

$$\leq \log C(d) + d \log 2 + \sum_{u=1}^{d} \log^+(\delta_u(D_y \phi_n))$$

we get that

$$I \leq \lim_{k \to \infty} \int_{\Omega_{k,l}} \sum_{i=1}^{m} \int_{\xi_{x_i}} \log C(d) + d \log 2 + \sum_{u=1}^{d} \log^+(\delta_u(D_y \phi_n)) d\mu(y) d\mathbb{P}$$

$$\leq \log C(d) + d \log 2 + \int_{\Omega_{k,l}} \int_{\mathbb{R}^d} \sum_{u=1}^{d} \log^+(\delta_u(D_y \phi_n)) d\mu(y) d\mathbb{P}.$$

6.3 THE MARGULIS-RUELLE INEQUALITY

To handle the terms II and III we will again estimate the number of elements of ξ_k that can intersect $\phi_n \xi_{x_i}$. To do this we put

$$L_k(n,\omega,i) := \sup_{z \in K_{\frac{1}{k}}(x_i)} ||D_z \phi_n|| \geq \sup_{z \in \xi_{x_i}} ||D_z \phi_n||$$

and observe that we have by the mean value theorem for $x, y \in \xi_{x_i}$ and arbitrary $\omega \in \Omega$

$$|\phi_n(x,\omega) - \phi_n(y,\omega)| \leq L_k(n,\omega,i)|x-y|$$

which implies

$$\phi_n \xi_{x_i} \subset \phi_n K_{\frac{1}{k}}(x_i) \subset K_{\frac{L_k(n,\omega,i)}{k}}(\phi_n(x_i,\omega)).$$

By Lemma 6.3.3 applied to $A := L_k(n,\omega,i)\mathrm{id}_{\mathbb{R}^d}$ we conclude

$$\#\{j : \xi_{x_j} \cap \phi_n(\cdot,\omega)\xi_{x_i} \neq \emptyset\} \leq C(d)(L_k(n,\omega,i) \vee 1)^d \tag{6.2}$$

which implies

$$II = -\lim_{k\to\infty} \int_\Omega \sum_{i=m+1}^\infty \mu(\xi_{x_i}) \sum_{j=1}^\infty \mu(\phi_n^{-1}\xi_{x_j}|\xi_{x_i}) \log \mu(\phi_n^{-1}\xi_{x_j}|\xi_{x_i}) d\mathbb{P}$$

$$\leq \lim_{k\to\infty} \int_\Omega \sum_{i=m+1}^\infty \mu(\xi_{x_i}) \log\left(C(d)(L_k(n,\omega,i) \vee 1)^d\right) d\mathbb{P}(\omega)$$

$$= \lim_{k\to\infty} \sum_{i=m+1}^\infty \mu(\xi_{x_i}) \int_\Omega \log C(d) + d\log^+ L_k(n,\omega,i) d\mathbb{P}(\omega)$$

$$= \lim_{k\to\infty} \sum_{i=m+1}^\infty \mu(\xi_{x_i}) \left(\log C(d) + d \int_\Omega \log^+ \left(L_k(n,\omega,1)\right) d\mathbb{P}(\omega)\right)$$

$$\leq \lim_{k\to\infty} \sum_{i=m+1}^\infty \mu(\xi_{x_i}) \left(\log C(d) + d \int_\Omega \log^+ \left(L_1(n,\omega,1)\right) d\mathbb{P}(\omega)\right)$$

$$= \mu(\mathbb{R}^d \setminus K_l(0)) \left(\log C(d) + d \int_\Omega \log^+ \left(L_1(n,\omega,1)\right) d\mathbb{P}(\omega)\right)$$

since by the rotation invariance of ϕ and (1.20) the distribution of $L_k(n,\cdot,i)$ does not depend on i and $L_k(n,\cdot,i)$ is decreasing in k. For the estimate (6.2) is valid for any i and

ω we may also use it to treat the term III.

$$III = -\lim_{k\to\infty} \int_{\Omega\setminus\Omega_{k,l}} \sum_{i=1}^{m} \mu(\xi_{x_i}) \sum_{j=1}^{\infty} \mu(\phi_n^{-1}\xi_{x_j}|\xi_{x_i}) \log \mu(\phi_n^{-1}\xi_{x_j}|\xi_{x_i}) d\mathbb{P}$$

$$\leq \lim_{k\to\infty} \sum_{i=1}^{m} \mu(\xi_{x_i}) \int_{\Omega\setminus\Omega_{k,l}} \log C(d) + d\log^+ L_k(n,\omega,i) d\mathbb{P}(\omega)$$

$$\leq \lim_{k\to\infty} \left(\mu(K_l(0)) \left(\log C(d) + d \int_{\Omega\setminus\Omega_{k,l}} \log^+ L_1(n,\omega,1) d\mathbb{P}(\omega) \right) \right).$$

Altogether we get that

$$nh_\mu(\phi) \leq \log C(d) + d\log 2 + \int_{\Omega_{k,l}} \int_{\mathbb{R}^d} \sum_{u=1}^{d} \log^+(\delta_u(D_y\phi_n)) d\mu(y) d\mathbb{P}$$
$$+ \lim_{k\to\infty} \mu(\mathbb{R}^d \setminus K_l(0)) \left(\log C(d) + d \int_{\Omega} \log^+ (L_1(n,\omega,1)) d\mathbb{P}(\omega) \right)$$
$$+ \lim_{k\to\infty} \left(\mu(K_l(0)) \left(\log C(d) + d \int_{\Omega\setminus\Omega_{k,l}} \log L_1(n,\omega,1) d\mathbb{P}(\omega) \right) \right)$$
$$\leq 2\log C(d) + d\log 2 + \int_{\Omega} \int_{\mathbb{R}^d} \sum_{u=1}^{d} \log^+(\delta_u(D_y\phi_n)) d\mu(y) d\mathbb{P}(\omega)$$
$$+ d \int_{\Omega} \log^+ (L_1(n,\omega,1)) d\mathbb{P}(\omega) \lim_{k\to\infty} \mu(\mathbb{R}^d \setminus K_l(0))$$
$$+ d \lim_{k\to\infty} \int_{\Omega\setminus\Omega_{k,l}} \log L_1(n,\omega,1) d\mathbb{P}(\omega). \tag{6.3}$$

Since $\int_\Omega \log^+ (L_1(n,\omega,1)) d\mathbb{P}(\omega)$ is finite which follows from [20, Theorem 2.2] the last term vanishes. If now we choose l in a way such that $\mu(\mathbb{R}^d \setminus K_l(0)) d \int_\Omega \log^+ L_1(n,\omega,1) d\mathbb{P}(\omega) \leq 42$ we may divide (6.3) by n to obtain

$$h_\mu(\phi) \leq \lim_{n\to\infty} \frac{1}{n} \int_\Omega \int_{\mathbb{R}^d} \sum_{u=1}^{d} \log^+(\delta_u(D_y\phi_n)) d\mu(y) d\mathbb{P}(\omega).$$

An application of Lemmas 6.4.1 and 6.4.2 (which we include in the next section for the convenience of the reader) now yields $h_\mu(\phi) \leq \lim_{n\to\infty} \frac{1}{n} \int_\Omega \int_{\mathbb{R}^d} |\widehat{D_y\phi_n}| d\mu(y) d\mathbb{P}(\omega) = \sum_{i=1}^{n} \lambda_i^+$ since for an IOUF the Lyapunov exponents are constant in space. $\square$

6.4 Exterior Powers, Tensor Algebras And SVDs

This part contains some facts from (multi-)linear algebra concerning the construction of certain algebras from a given finite-dimensional vector space and the formulation of how this is applied to prove Ruelle's inequality. We only give the definitions and statements for ease of reference and omit the proofs. Let V be a finite-dimensional real vector space, $n := \dim(V)$ and denote by

$$V^{\otimes} := \bigoplus_{k=0}^{\infty} \bigotimes_{i=1}^{k} V^{\otimes i}$$

the associated (contravariant) tensor algebra. Let further $\mathcal{I} := \mathcal{I}(v \otimes v : v \in V)$ be the ideal generated by elements of the form $v \otimes v$. The factor algebra

$$V^{\wedge} := V^{\otimes}/\mathcal{I}$$

is the outer algebra of V and it can be easily shown to admitt the following decomposition.

$$V^{\wedge} = \bigoplus_{k=0}^{n} V^{\wedge k} \text{ with } V^{\wedge k} = V^{\otimes k} \cap V^{\wedge} \text{ and } \dim(V^{\wedge k}) = \frac{n!}{k!(n-k)!}$$

We now take a vector space endomorphism $T\colon V \to V$ and seek to define $T^{\wedge k}$. One might consider $\mathrm{Hom}(V)$ as a vector space and perform the above for the vector T. This would immediatly lead to $T^{\wedge k} = 0$ and hence is not the right way to do this. For „going from a vector space to its tensor algebra" is a covariant functor (as we will see) it has to transform the morphisms of V into morphisms of the tensor algebra and hence (mod $\mathcal{I}$) to morphisms of the exterior algebra. It turns out that T can be extended to an algebra endomorphism $T^{\otimes}$ on $V^{\otimes}$ in unique manner. Since $T^{\otimes}$ preserves $V^{\otimes k}$ we can get $T^{\wedge k}$ as the restriction of $T^{\otimes}$ to $V^{\wedge k}$. Exterior powers $T^{\wedge k}$ play an important role in the multiplicative ergodic theorem since their singular values link generically to the Lyapunov exponents of the systems. Before stating the main results we introduce the shorthand

$$T^{\wedge}\colon V^{\wedge} \to V^{\wedge}, T^{\wedge} := \mathrm{id}_{\mathbb{R}} \oplus T \oplus T^{\wedge 2} \ldots T^{\wedge n}.$$

Lemma 6.4.1. *Let ϕ be an IOUF with corresponding Lyapunov exponents $\lambda_1, \ldots, \lambda_d$ Then we have*

$$\prod_{u=1}^{d}(1 \vee \delta_u(D_y\phi_n)) \leq \log|\widehat{D_y\phi_n(y,\omega)}|.$$

Proof: This is shown in a much more general context in [1, II.(2.6)]. □

Lemma 6.4.2. *Let ϕ be an IOUF with corresponding Lyapunov exponents $\lambda_1, \ldots, \lambda_d$. Then we have*

$$\lim_{n\to\infty} \frac{1}{n} \int_\Omega \int_{\mathbb{R}^d} \log|\widehat{D_y\phi_n(y,\omega)}| d\mu(y) d\mathbb{P}(\omega) = \int_{\mathbb{R}^d} \sum_{i=1}^{d} \lambda_i^+(y) d\mu(y).$$

Proof: This lemma is valid in the context of RDS on manifolds as shown in [1, Proposition II.3.2]. See also [49]. □

Chapter 7

Asymptotic Growth Of Spatial Derivatives Of Isotropic Flows

It is known from the multiplicative ergodic theorem (cf. Theorem 1.2.9) that the norm of the derivative of certain stochastic flows at a previously fixed point grows exponentially fast in time as the flows evolve. We prove that this is also true if one takes the supremum over a bounded set of initial points. We give an explicit bound for the exponential growth rate which is far different from the lower bound coming from the Multiplicative Ergodic Theorem. We start with an introduction and recall some preliminary lemmas from the literature in the first section, give the main result in the second one and devote the remaining part of the chapter to its proof. In the whole chapter we will use ϕ to denote an IOUF or a RIF with covariance tensor b and drift c or an IBF with covariance tensor b. Since the cases of IOUFs, RIFs and IBFs can be covered with completely the same computations we do not need to distinguish between the three cases here.

7.1 Introduction And Preliminaries

The evolution of the diameter of a bounded set under the evolution of a stochastic flow has been studied since the 1990's and we treated the special case of a two-dimensional IBF in Chapter 4. Of course the considered diameter links to the supremum of $|\phi_t(x)|$ with x ranging over a subset of $\mathbb{R}^d$ and hence this supremum also grows linearly in time. In the following we will consider the case where the flow is replaced by its spatial derivative. We emphasize that we consider the asymptotics in time (the spatial asymptotics for a fixed time horizon have been considered in [20] in a very general setting and in [57] in the particular case treated here - cf. Chapter 2). If the flow has a positive top exponent it is known that the growth is at least exponentially fast which is then true even for a singleton (this follows directly from the multiplicative ergodic theorem). We will show in the case of an isotropic flow that $\sup_x |\log \|D\phi_t(x)\||$ grows at most linearly in time t where the supremum is taken over x in a bounded subset of $\mathbb{R}^d$ no matter what the top Lyapunov exponent is. This shows that the growth of the norm of the derivative is indeed at most exponentially fast but also gives some insight into the distance of $D\phi_t(x)$ to singularity, which might be of interest especially if the top exponent is negative. Exponential bounds on the growth of spatial derivatives play a role in the proof of Pesin's formula for stochastic flows (see [37]). We will also give a new proof of the fact that the diameter grows at most linearly in time from this which is originally due to T. Lyons (but there are much simpler proofs known for this - see the references given before). Despite the fact that we can come up with an upper bound for the exponential growth rate we make no claims about its optimality (and we conjecture that our bound is far from optimal). The first lemmas collect severals estimates we will have to use in the sequel which are obtained from straightforward calculus and are stated for the ease of reference and to make later computations more readable.

Lemma 7.1.1 (dirty estimates).
The following assertions hold.

1. *For $t > 0$ the cumulative distribution function of the normal Φ satisfies $1 - \Phi(t) \leq e^{-\frac{1}{2}t^2}$.*

2. *For $t \geq 1$ Stirlings formula yields for the Gamma function $\Gamma\left(\frac{t+1}{2}\right)^{\frac{1}{t}} \leq \sqrt{2\pi} e^{-\frac{5}{12}} \sqrt{t+1}$.*

3. *For $t \geq 0$ we have $t \leq e^{\frac{t}{e}}$.*

4. *For $t \geq 0$ we have $\sqrt{t} \leq t + \frac{1}{4}$.*

5. *For $t \geq 1$ we have $\sqrt{\log t} \leq t^{\frac{1}{2e}}$.*

Proof: One can derive these straightforward from undergraduate calculus. Note that the estimate $\Gamma(x) \leq \sqrt{2\pi} x^{1-\frac{1}{2}} e^{\frac{1}{12x}}$ for $x > 0$ which can be found e.g. in [26, page 359] is useful. □

7.1 Introduction and Preliminaries

Lemma 7.1.2 (a lemma on real functions).
Let $f, g : [0, \infty) \to [0, \infty)$ be increasing functions that are differentiable on $(0, \infty)$. Let further f be convex and g be concave. If we have for some $t > 0$ that $f(t) \geq g(t)$ and $f'(t) \geq g'(t)$ then we have for all $s \geq t$ that $f(s) \geq g(s)$.

Proof: This is an elementary undergraduate exercise. $\square$

The following result is the main tool that allows for the estimation of suprema of the derivatives.

Theorem 7.1.3 (Chaining Growth Theorem).
Assume $\psi : [0, \infty) \times \mathbb{R}^d \times \Omega \to \mathbb{R}^d$ is a continuous random field with the following properties.

1. There exist $A > 0$ and $B \geq 0$ such that for each $k > 0$ and each bounded set $S \subset \mathbb{R}^d$ we have
$$\limsup_{T \to \infty} \frac{1}{T} \log \sup_{x \in S} \mathbb{P}\left[\sup_{0 \leq t \leq T} |\psi_t(x)| > kT\right] \leq -\frac{(k-B)_+^2}{2A^2}$$
where $r_+ = r \vee 0$ denotes the positive part of $r \in \mathbb{R}$.

2. There exist $\Lambda \geq 0, \sigma > 0, q_0 \geq 1$ and $\bar{c} > 0$ such that for each $x, y \in \mathbb{R}^d$, $T > 0$ and even $q \geq q_0$ we have
$$\mathbb{E}\left[\sup_{0 \leq t \leq T} |\psi_t(x) - \psi_t(y)|^q\right]^{1/q} \leq \bar{c} |x - y| e^{(\Lambda + \frac{1}{2} q \sigma^2) T}.$$

$\bar{c}$ may depend on q and d but neither on $|x - y|$ nor on T.

Let Ξ be a compact subset of $\mathbb{R}^d$ with box (or upper entropy dimension: see [52, page 19] for a definition; just note that a closed ball in $\mathbb{R}^d$ has box dimension d) dimension $\Delta > 0$. Then
$$\limsup_{T \to \infty} \left(\sup_{0 \leq t \leq T} \sup_{x \in \Xi} \frac{1}{T} |\psi_t(x)|\right) \leq K \quad \text{a.s.}$$
where
$$K := \begin{cases} B + A\sqrt{2\Delta \left(\Lambda + \sigma^2 \Delta + \sqrt{\sigma^4 \Delta^2 + 2\Delta \Lambda \sigma^2}\right)} & : \text{if } \Lambda \geq \Lambda_0 \\ B + A\sqrt{2\Delta \frac{d}{d-\Delta} \left(\Lambda + \frac{1}{2} \sigma^2 d\right)} & : \text{otherwise} \end{cases}$$
for
$$\Lambda_0 := \frac{\sigma^2 d}{\Delta} \left(\frac{d}{2} - \Delta\right).$$

Proof: [52, Theorem 5.1]. Observe that the change of „$q \geq 1$" to „even $q \geq q_0$" does not alter the statement at all, because the assumptions above guarantee the assumptions for $q \geq 1$ with a change only in the value of $\bar{c}$. The fact that we allow $\bar{c}$ to depend on q does not play any role because the proof in [52] is perfectly valid with q-depending $\bar{c}$. □

We finally recall the Burkholder-Davies-Gundy inequality because we will use the precise asymptotics for the constants appearing therein.

Lemma 7.1.4 (Burkholder-Davies-Gundy inequality).
For $q \geq 1$ there exists a constant C_q such that for every continuous local martingale $(M_t)_{t \geq 0}$ with $M_0 = 0$ we have
$$\mathbb{E}\left[\sup_{0 \leq t \leq T} M_t^q\right] \leq C_q \mathbb{E}\left[\langle M \rangle_T^{\frac{q}{2}}\right].$$

In fact it is known that for even q the optimal $C_q^{\frac{1}{q}}$ is the largest positive zero of the Hermite polynomial of order $2q$ and can be estimated via $C_q^{\frac{1}{q}} \leq k\sqrt{4q+1}$ for some constant $k > 0$. If one restricts oneself to large values of q one may choose k to be at most $\sqrt{2}$.

Proof: [11] for the optimality of the constant and [47] for the estimate (and also for a more precise statement about the asymptotics of the maximal zeros of the Hermite polynomials). □

7.2 The Main Result

We are now ready to state the main result.

Theorem 7.2.1 (exponential growth of spatial derivatives).
Let $(\phi_{s,t} : 0 \leq s \leq t < \infty)$ be an IOUF as in Definition 1.4.1, an IBF as in Definition 1.3.3 or a RIF as in Definition 1.5.1. In all cases we denote the drift by c and the covariance tensor by b (notation as before). Let Ξ be a compact subset of $\mathbb{R}^d$ with box dimension $\Delta > 0$. Then
$$\limsup_{T \to \infty} \left(\sup_{0 \leq t \leq T} \sup_{x \in \Xi} \frac{1}{T} |\log \|D\phi_t(x)\|| \right) \leq K \text{ a.s.}$$

where
$$K := \begin{cases} B + A\sqrt{2\Delta \left(\Lambda + \sigma^2 \Delta + \sqrt{\sigma^4 \Delta^2 + 2\Delta\Lambda\sigma^2}\right)} & : \text{ if } \Lambda \geq \Lambda_0 \\ B + A\sqrt{2\Delta \frac{d}{d-\Delta} \left(\Lambda + \frac{1}{2}\sigma^2 d\right)} & : \text{ otherwise} \end{cases}$$

for

$$\Lambda_0 := \frac{\sigma^2 d}{\Delta}\left(\frac{d}{2}-\Delta\right), \qquad A := \sqrt{\beta_L},$$

$$B := 0 \vee \frac{(d-1)\beta_N + \beta_L - 2c}{2} \vee \frac{\beta_L - (d-1)\beta_N + 2c}{2} = \frac{\beta_L}{2} + \frac{|(d-1)\beta_N - 2c|}{2},$$

$$\Lambda := \Lambda_1 \vee \Lambda_2 \vee \Lambda_6, \qquad \sigma := \sigma_1 \vee \sigma_2 \vee \sigma_6.$$

The Λ_i and σ_i are constants that depend on b and d and will be specified later.

Remark: The formula for the constant K above looks like the upper bound one can have for the expansion speed of the flow itself (cf. [52]). This is due to the fact that Theorem 7.1.3 is taylored to give good estimates in that case. The differences enter via different values for A, B, Λ and σ^2.

Proof: This follows directly from Theorem 7.1.3 applied to

$$\psi : [0,\infty) \times \mathbb{R}^d \times \Omega \to \mathbb{R}; \psi_t(x,\omega) := \log \|D\phi_t(x)\|$$

if we can verify the following lemmas. In the entire chapter we will only use ψ in the meaning given above from now on. Note that we choose the matrix norm be the Frobenius norm $\|(a_{i,j})_{1 \le i,j \le d}\| := \left(\sum_{i,j} a_{i,j}^2\right)^{1/2}$ for its computational simplicity although the special choice of a norm is irrelevant because of their equivalence.

Lemma 7.2.2 (condition on the one-point motion).
We have for each bounded $S \subset \mathbb{R}^d$ that

$$\limsup_{T\to\infty} \frac{1}{T} \log \sup_{x \in S} \mathbb{P}\left[\sup_{0 \le t \le T} |\psi_t(x)| > kT\right] \le -\frac{(k-B)_+^2}{2A^2}$$

for A and B as given in Theorem 7.2.1.

Lemma 7.2.3 (condition on the two-point motion).
We have for each $x,y \in \mathbb{R}^d$, $T > 0$ and even $q \ge q_0 := \frac{4\sqrt{\beta_L}[2(d-1)\beta_N - c - 2\beta_L]}{128\beta_L} \vee 3$ that

$$\mathbb{E}\left[\sup_{0 \le t \le T} |\psi_t(x) - \psi_t(y)|^q\right]^{1/q} \le \bar{c}\sqrt{q}e^{(\Lambda + \frac{1}{2}q\sigma^2)T}$$

for Λ and σ as given in Theorem 7.2.1 and $\bar{c} := \bar{c}_1 + \bar{c}_2 + \bar{c}_6$. The $\bar{c}_i$ are constants that depend on b and d and will be specified later.

The proofs of these lemmas will be given in the next sections. Observe that $\bar{c}$ does not enter into the constant K, so we do not need to pay attention to get a small value for it.

7.3 Proof Of Lemma 7.2.2: The One-Point Condition

Before proving Lemma 7.2.2 we will need to establish some facts on $\|D\phi_t(x)\|^2$.

Lemma 7.3.1 (SDE for $\|D\phi_t(x)\|^2$).
Let $x \in \mathbb{R}^d$ and put

$$M_t := 2 \int_0^t \sum_{i,j,k} \partial_k M^i(ds, x_s) \frac{\partial_j \phi_s^k(x) \, \partial_j \phi_s^i(x)}{\|D\phi_s(x)\|^2}.$$

Then we have the following.

1. $(M_t)_{t \geq 0}$ is a continuous local martingale that a.s. satifies for $t \geq 0$ that

$$\langle M \rangle_t = 2(\beta_L - \beta_N)t + 2(\beta_L + \beta_N) \int_0^t \sum_{i,j,k,m} \frac{\partial_j \phi_s^k(x) \, \partial_j \phi_s^i(x) \, \partial_m \phi_s^k(x) \, \partial_m \phi_s^i(x)}{\|D\phi_s(x)\|^4} ds \leq 4\beta_L t$$

and hence is a true martingale.

2. $\|D\phi_t(x)\|^2$ solves the SDE

$$\begin{aligned}\|D\phi_t(x)\|^2 &= d + 2 \int_0^t \sum_{i,j,k} \partial_k M^i(ds, x_s) \, \partial_j \phi_s^k(x) \, \partial_j \phi_s^i(x) \\ &\quad + [(d-1)\beta_N + \beta_L - 2c] \int_0^t \|D\phi_s(x)\|^2 \, ds \\ &= d + \int_0^t \|D\phi_s(x)\|^2 \, dM_s + [(d-1)\beta_N + \beta_L - 2c] \int_0^t \|D\phi_s(x)\|^2 \, ds.\end{aligned}$$

7.3 Proof Of Lemma 7.2.2: The One-Point Condition

3. The quadratic variation $\left\langle \|D\phi_{\cdot}(x)\|^2 \right\rangle_t$ satisfies

$$\left\langle \|D\phi_{\cdot}(x)\|^2 \right\rangle_t = 2(\beta_L - \beta_N) \int_0^t \|D\phi_s(x)\|^4 \, ds$$
$$+ 2(\beta_N + \beta_L) \int_0^t \sum_{i,j,k,m} \partial_j \phi_s^k(x) \, \partial_j \phi_s^i(x) \, \partial_m \phi_s^k(x) \, \partial_m \phi_s^i(x) \, ds.$$

4. $\psi_t(x) = \log \|D\phi_t(x)\|$ solves the SDE

$$\psi_t(x) = \frac{1}{2} \log d + \frac{1}{2} M_t + \left[\frac{(d-1)\beta_N + \beta_L}{2} - c \right] t - \frac{1}{4} \langle M \rangle_t.$$

Proof: Since we have by Theorem 1.3.6 and Lemma 1.4.2 that

$$\partial_j \phi_t^i(x) = \delta_{ij} + \int_0^t \sum_k \partial_j \phi_s^k(x) \, \partial_k M^i(ds, x_s) - c \int_0^t \partial_j \phi_t^i(x) \, ds$$

we also get that Itô's formula implies for $\|D\phi_t(x)\|^2 = \sum_{i,j}(\partial_j \phi_t^i(x))^2$ that

$$\|D\phi_t(x)\|^2 = d + 2 \sum_{i,j} \int_0^t \partial_j \phi_s^i(x) \, d\partial_j \phi_s^i(x) + \sum_{i,j} \left\langle \partial_j \phi_{\cdot}^i(x) \right\rangle_t.$$

By Theorem 1.3.6 and Lemma 1.4.2 this is equal to

$$d + 2 \sum_{i,j,k} \int_0^t \partial_j \phi_s^i(x) \, \partial_k M^i(ds, x_s) \, \partial_j \phi_s^k(x) - 2c \sum_{i,j} \int_0^t (\partial_j \phi_s^i(x))^2 ds$$
$$+ \sum_{i,j} \int_0^t \sum_{k,l} \partial_j \phi_s^l(x) \, \partial_j \phi_s^k(x) \, d\left\langle \partial_l M^i(d\cdot, x.), \partial_k M^i(d\cdot, x.) \right\rangle_s$$
$$= d + 2 \sum_{i,j,k} \int_0^t \partial_k M^i(ds, x_s) \, \partial_j \phi_s^i(x) \, \partial_j \phi_s^k(x) - 2c \int_0^t \|D\phi_s(x)\|^2 \, ds$$
$$- \sum_{i,j,k,l} \int_0^t \partial_j \phi_s^l(x) \, \partial_j \phi_s^k(x) \, \partial_k \partial_l b^{i,i}(0) \, ds. \tag{7.1}$$

Since we have by Lemma 1.3.2

$$\sum_{i,j,k,l} \int_0^t \partial_j \phi_s^l(x) \partial_j \phi_s^k(x) \partial_k \partial_l b^{i,i}(0) \, ds$$

$$= \int_0^t \sum_{i,j,k,l} \partial_j \phi_s^l(x) \partial_j \phi_s^k(x) [(\beta_N - \beta_L)\delta_{ki}\delta_{li} - \beta_N \delta_{kl}] \, ds$$

$$= \int_0^t \sum_{i,j} (\beta_N - \beta_L)(\partial_j \phi_s^i(x))^2 ds - d\beta_N \int_0^t \sum_{j,k} (\partial_j \phi_s^k(x))^2 ds$$

$$= -[\beta_L + (d-1)\beta_N] \int_0^t \|D\phi_s(x)\|^2 \, ds$$

we also get from (7.1) that $\|D\phi_t(x)\|^2$ equals

$$d + 2 \sum_{i,j,k} \int_0^t \partial_k M^i(ds, x_s) \partial_j \phi_s^i(x) \partial_j \phi_s^k(x)$$

$$- 2c \int_0^t \|D\phi_s(x)\|^2 \, ds + [\beta_L + (d-1)\beta_N] \int_0^t \|D\phi_s(x)\|^2 \, ds$$

$$= d + 2 \sum_{i,j,k} \int_0^t \|D\phi_s(x)\|^2 \partial_k M^i(ds, x_s) \frac{\partial_j \phi_s^i(x) \partial_j \phi_s^k(x)}{\|D\phi_s(x)\|^2}$$

$$+ [\beta_L + (d-1)\beta_N - 2c] \int_0^t \|D\phi_s(x)\|^2 \, ds$$

$$= d + 2 \int_0^t \|D\phi_s(x)\|^2 \, dM_s + [\beta_L + (d-1)\beta_N - 2c] \int_0^t \|D\phi_s(x)\|^2 \, ds.$$

This proves assertion 2.. 4. follows from this and Itô's formula since $\psi_t(x) = \frac{1}{2} \log \left(\|D\phi_t(x)\|^2\right)$. To prove 1. and 3. we observe that by (1.9) and Lemma 1.3.2 we have

$$\langle M \rangle_t$$

$$= \left\langle 2 \int_0^\cdot \sum_{i,j,k} \partial_k M^i(ds, x_s) \frac{\partial_j \phi_s^k(x) \partial_j \phi_s^i(x)}{\|D\phi_s(x)\|^2}, 2 \int_0^\cdot \sum_{l,m,n} \partial_n M^l(ds, x_s) \frac{\partial_m \phi_s^n(x) \partial_m \phi_s^l(x)}{\|D\phi_s(x)\|^2} \right\rangle$$

$$= -4 \sum_{i,j,k,l,m,n} \int_0^t \frac{\partial_j \phi_s^k(x) \partial_j \phi_s^i(x) \partial_m \phi_s^n(x) \partial_m \phi_s^l(x)}{\|D\phi_s(x)\|^4} \partial_k \partial_n b^{i,l}(0) \, ds$$

$$= -4 \sum_{i,j,k,l,m,n} \int_0^t \frac{\partial_j \phi_s^k(x) \partial_j \phi_s^i(x) \partial_m \phi_s^n(x) \partial_m \phi_s^l(x)}{\|D\phi_s(x)\|^4}$$

$$\left[\frac{1}{2}(\beta_N - \beta_L)(\delta_{ki}\delta_{nl} + \delta_{kl}\delta_{ni}) - \beta_N \delta_{kn}\delta_{il}\right] ds$$

7.3 Proof Of Lemma 7.2.2: The One-Point Condition

$$= 2(\beta_L - \beta_N) \int_0^t \sum_{i,j,l,m} \frac{(\partial_j \phi_s^i(x))^2 (\partial_m \phi_s^l(x))^2}{\|D\phi_s(x)\|^4} ds$$

$$+ 2(\beta_L - \beta_N) \int_0^t \sum_{i,j,k,m} \frac{\partial_j \phi_s^k(x) \partial_j \phi_s^i(x) \partial_m \phi_s^i(x) \partial_m \phi_s^k(x)}{\|D\phi_s(x)\|^4} ds$$

$$+ 4\beta_N \int_0^t \sum_{i,j,k,m} \frac{\partial_j \phi_s^k(x) \partial_j \phi_s^i(x) \partial_m \phi_s^k(x) \partial_m \phi_s^i(x)}{\|D\phi_s(x)\|^4} ds$$

$$= 2(\beta_L - \beta_N)t + 2(\beta_L + \beta_N) \int_0^t \sum_{i,j,k,m} \frac{\partial_j \phi_s^k(x) \partial_j \phi_s^i(x) \partial_m \phi_s^k(x) \partial_m \phi_s^i(x)}{\|D\phi_s(x)\|^4} ds.$$

This together with 2. proves 3. and 1. also follows from this and the next proposition. □

Proposition 7.3.2 (a simple estimate).
We have $\left| \sum_{i,j,k,m} \frac{\partial_j \phi_s^k(x) \partial_j \phi_s^i(x) \partial_m \phi_s^k(x) \partial_m \phi_s^i(x)}{\|D\phi_s(x)\|^4} \right| \leq 1.$

Proof: Applying the triangle inequality and Schwarz' inequality we get

$$\left| \sum_{i,j,k,m} \frac{\partial_j \phi_s^k(x) \partial_j \phi_s^i(x) \partial_m \phi_s^k(x) \partial_m \phi_s^i(x)}{\|D\phi_s(x)\|^4} \right|$$

$$\leq \sum_{i,k} \left| \sum_j \frac{\partial_j \phi_s^k(x) \partial_j \phi_s^i(x)}{\|D\phi_s(x)\|^2} \right| \left| \sum_m \frac{\partial_m \phi_s^k(x) \partial_m \phi_s^i(x)}{\|D\phi_s(x)\|^2} \right|$$

$$\leq \sum_{i,k} \sqrt{\sum_j \left(\frac{\partial_j \phi_s^k(x)}{\|D\phi_s(x)\|} \right)^2} \sqrt{\sum_j \left(\frac{\partial_j \phi_s^i(x)}{\|D\phi_s(x)\|} \right)^2} \sqrt{\sum_m \left(\frac{\partial_m \phi_s^k(x)}{\|D\phi_s(x)\|} \right)^2} \sqrt{\sum_m \left(\frac{\partial_m \phi_s^i(x)}{\|D\phi_s(x)\|} \right)^2}$$

$$= \sum_{i,k} \sum_j \left(\frac{\partial_j \phi_s^k(x)}{\|D\phi_s(x)\|} \right)^2 \sum_m \left(\frac{\partial_m \phi_s^i(x)}{\|D\phi_s(x)\|} \right)^2 = \sum_{j,k} \left(\frac{\partial_j \phi_s^k(x)}{\|D\phi_s(x)\|} \right)^2 \sum_{i,m} \left(\frac{\partial_m \phi_s^i(x)}{\|D\phi_s(x)\|} \right)^2 = 1.$$

□

Now we can turn to the proof of Lemma 7.2.2.
Since we can write $M_t = W_{\langle M \rangle_t}$ for a standard Brownian motion $(W_t)_{t \geq 0}$ we get with Lemma 7.3.1

$$\psi_t(x) = \frac{1}{2} \log d + \left[\frac{(d-1)\beta_N + \beta_L}{2} - c \right] t + \frac{1}{2} \left(W_{\langle M \rangle_t} - \frac{1}{2} \langle M \rangle_t \right)$$

$$\leq \frac{1}{2} \log d + \left[\frac{(d-1)\beta_N + \beta_L}{2} - c \right] t + \frac{1}{2} \sup_{0 \leq s \leq 4\beta_L t} \left(W_s - \frac{1}{2} s \right)$$

$$= \frac{1}{2}\log d + \left[\frac{(d-1)\beta_N + \beta_L}{2} - c\right]t + \frac{1}{2}\sup_{0\leq s\leq 1}\left(W_{4\beta_L st} - \frac{4\beta_L st}{2}\right)$$

$$= \frac{1}{2}\log d + \left[\frac{(d-1)\beta_N + \beta_L}{2} - c\right]t + \sqrt{\beta_L}t \sup_{0\leq s\leq 1}\left(\frac{W_{4\beta_L ts}}{2\sqrt{\beta_L}t} - \sqrt{\beta_L}ts\right)$$

$$\stackrel{d}{=} \frac{1}{2}\log d + \left[\frac{(d-1)\beta_N + \beta_L}{2} - c\right]t + \sqrt{\beta_L}t \sup_{0\leq s\leq 1}\left(W_s - \sqrt{\beta_L}ts\right) \qquad (7.2)$$

where the latter means equality in distribution. Therefore we get for any $k > 0$

$$\mathbb{P}\left[\sup_{0\leq t\leq T} \psi_t(x) \geq kT\right] \qquad (7.3)$$

$$\leq \mathbb{P}\left[\frac{\log d}{2} + \sup_{0\leq t\leq T}\left\{\left[\frac{(d-1)\beta_N + \beta_L}{2} - c\right]t + \sqrt{\beta_L}t \sup_{0\leq s\leq 1}\left(W_s - \sqrt{\beta_L}ts\right)\right\} \geq kT\right]$$

$$\leq \mathbb{P}\left[\sup_{0\leq t\leq T}\left\{\left[\frac{(d-1)\beta_N + \beta_L - 2c}{2\sqrt{\beta_L}}\right]_+ t + \sqrt{t} \sup_{0\leq s\leq 1}\left(W_s - \sqrt{\beta_L}ts\right)\right\} \geq \frac{k}{\sqrt{\beta_L}}T - \frac{\log d}{2\sqrt{\beta_L}}\right] =: I.$$

Here we distinguish between two cases to treat (7.3). If $(d-1)\beta_N + \beta_L - 2c \leq 0$ then we immediately get

$$I \leq \mathbb{P}\left[\sqrt{T} \sup_{0\leq s\leq 1} W_s \geq \frac{k}{\sqrt{\beta_L}}T - \frac{\log d}{2\sqrt{\beta_L}}\right] = \mathbb{P}\left[\sup_{0\leq s\leq 1} W_s \geq \frac{k}{\sqrt{\beta_L}}\sqrt{T} - \frac{\log d}{2\sqrt{\beta_L T}}\right]$$

$$= 2\left[1 - \Phi\left(\frac{k\sqrt{T}}{\sqrt{\beta_L}} - \frac{\log d}{2\sqrt{\beta_L T}}\right)\right] \leq 2e^{-\frac{1}{2}\left(\frac{k^2 T}{\beta_L} - \frac{k\log d}{\beta_L} + \frac{(\log d)^2}{4\beta_L T}\right)} \qquad (7.4)$$

which gives in this case $\limsup_{T\to\infty} \frac{1}{T}\log \mathbb{P}\left[\sup_{0\leq t\leq T}\psi_t(x) \geq kT\right] \leq -\frac{1}{2\beta_L}k^2 = -\frac{1}{2\beta_L}k_+^2$. Now let $[(d-1)\beta_N + \beta_L] - 2c > 0$ and observe

$$I \leq \mathbb{P}\left[\sup_{0\leq t\leq T}\left\{\left[\frac{(d-1)\beta_N + \beta_L - 2c}{2\sqrt{\beta_L}}\right]t + \sqrt{t} \sup_{0\leq s\leq 1} W_s\right\} \geq \frac{k}{\sqrt{\beta_L}}T - \frac{\log d}{2\sqrt{\beta_L}}\right]$$

$$= \mathbb{P}\left[\sup_{0\leq s\leq 1} W_s \geq \left[\frac{k}{\sqrt{\beta_L}} + \frac{2c - (d-1)\beta_N - \beta_L}{2\sqrt{\beta_L}}\right]\sqrt{T} - \frac{\log d}{2\sqrt{\beta_L T}}\right]$$

7.3 Proof Of Lemma 7.2.2: The One-Point Condition

which leads (as in (7.4)) to

$$\limsup_{T\to\infty} \frac{1}{T} \log \mathbb{P}\left[\sup_{0\le t\le T} \psi_t(x) \ge kT\right] \le -\frac{1}{2}\left[\frac{k}{\sqrt{\beta_L}} + \frac{2c - (d-1)\beta_N - \beta_L}{2\sqrt{\beta_L}}\right]^2$$
$$\le -\frac{1}{2\beta_L}\left[k - \frac{(d-1)\beta_N + \beta_L - 2c}{2}\right]_+^2.$$

We now only have to exclude the possibility that the modulus of the logarithm in $\psi_t(x)$ might become large due to a very small $\|D\phi_t(x)\|$. Observe, that (7.2) implies

$$\psi_t(x) = \frac{1}{2}\log d + \left[\frac{(d-1)\beta_N + \beta_L}{2} - c\right]t + \frac{1}{2}\left(W_{\langle M\rangle_t} - \frac{1}{2}\langle M\rangle_t\right) \qquad (7.5)$$
$$\ge \left[\frac{(d-1)\beta_N + \beta_L}{2} - c\right]t + \frac{1}{2}W_{\langle M\rangle_t} - \beta_L t = \left[\frac{(d-1)\beta_N - \beta_L - 2c}{2}\right]t + \frac{1}{2}W_{\langle M\rangle_t}.$$

If $(d-1)\beta_N - \beta_L - 2c \ge 0$ we may thus estimate

$$\mathbb{P}\left[\inf_{0\le t\le T}\psi_t(x) \le -kT\right] \le \mathbb{P}\left[\frac{1}{2}\inf_{0\le t\le 4\beta_L T} W_t \le -kT\right]$$
$$= \mathbb{P}\left[\frac{1}{2}\sup_{0\le t\le 4\beta_L T} W_t \ge kT\right] = \mathbb{P}\left[\sup_{0\le t\le 1} W_t \ge \frac{k}{\sqrt{\beta_L}}\sqrt{T}\right]$$

which implies (see (7.4)) that $\limsup_{T\to\infty}\frac{1}{T}\log \mathbb{P}\left[\inf_{0\le t\le T}\psi_t(x)\le -kT\right] \le -\frac{1}{2\beta_L}k_+^2$. In the remaining case $(d-1)\beta_N - \beta_L - 2c \le 0$ we observe that (7.5) yields

$$\mathbb{P}\left[\inf_{0\le t\le T}\psi_t(x) \le -kT\right] \le \mathbb{P}\left[\frac{1}{2}\inf_{0\le t\le 4\beta_L T} W_t \le \left[\frac{-(d-1)\beta_N + \beta_L + 2c}{2} - k\right]T\right]$$
$$= \mathbb{P}\left[\sup_{0\le t\le 1} W_t \ge \left[\frac{(d-1)\beta_N - \beta_L - 2c}{2\sqrt{\beta_L}} + \frac{k}{\sqrt{\beta_L}}\right]\sqrt{T}\right]$$

and so

$$\limsup_{T\to\infty}\frac{1}{T}\log\mathbb{P}\left[\inf_{0\le t\le T}\psi_t(x)\le -kT\right] \le -\frac{1}{2\beta_L}\left[\frac{(d-1)\beta_N - \beta_L - 2c}{2} + k\right]^2$$
$$\le -\frac{1}{2\beta_L}\left[k - \frac{\beta_L - (d-1)\beta_N + 2c}{2}\right]_+^2.$$

This completes the proof of Lemma 7.2.2. □

7.4 Proof Of Lemma 7.2.3: The Two-Point Condition

7.4.1 General Estimates And Preparation

We now turn to the Proof of Lemma 7.2.3. Observe that we have by Lemma 7.3.1

$$\psi_t(x) - \psi_t(y) = \int_0^t \sum_{i,j,k} \partial_k M^i (ds, x_s) \frac{\partial_j \phi_s^k(x) \partial_j \phi_s^i(x)}{\|D\phi_s(x)\|^2} - \partial_k M^i(ds, y_s) \frac{\partial_j \phi_s^k(y) \partial_j \phi_s^i(y)}{\|D\phi_s(y)\|^2}$$

$$+ \frac{(\beta_L + \beta_N)}{2} \int_0^t \sum_{i,j,k,m} \frac{\partial_j \phi_s^k(y) \partial_j \phi_s^i(y) \partial_m \phi_s^k(y) \partial_m \phi_s^i(y)}{\|D\phi_s(y)\|^4}$$

$$- \frac{\partial_j \phi_s^k(x) \partial_j \phi_s^i(x) \partial_m \phi_s^k(x) \partial_m \phi_s^i(x)}{\|D\phi_s(x)\|^4} ds$$

$$=: \tilde{M}_t + A_t.$$

To further analyze the latter we first prove the following lemma.

Lemma 7.4.1 (general estimates for A_t and $\tilde{M}_t$).
With $\tilde{M}_t$ and A_t defined as above we have the following.

1. *A.s. we have for all $t \geq 0$ that $A_t \leq (\beta_N + \beta_L)t$.*

2. *The quadratic variation satisfies*

$$\langle \tilde{M} \rangle_t$$

$$= \frac{\beta_L + \beta_N}{2} \int_0^t \sum_{i,k} \left(\sum_j \frac{\partial_j \phi_s^i(x) \partial_j \phi_s^k(x)}{\|D\phi_s(x)\|^2} - \frac{\partial_j \phi_s^i(y) \partial_j \phi_s^k(y)}{\|D\phi_s(y)\|^2} \right)^2 ds$$

$$+ 2 \sum_{i,j,k,l,m,n} \int_0^t \frac{\partial_j \phi_s^i(x) \partial_j \phi_s^k(x) \partial_m \phi_s^l(y) \partial_m \phi_s^n(y)}{\|D\phi_s(x)\|^2 \|D\phi_s(y)\|^2} \left[\partial_k \partial_n b^{i,l}(x_s - y_s) - \partial_k \partial_n b^{i,l}(0) \right] ds.$$

3. $\langle \tilde{M} \rangle_t \leq \tilde{c} t$ *for* $\tilde{c} := 2\beta_L + 2d^6 \max_{i,l,k,n} \sup_{z \in \mathbb{R}^d} \partial_k \partial_n b^{i,l}(z)$.

7.4 Proof Of Lemma 7.2.3: The Two-Point Condition

Proof: 1. is clear by definition of A_t and Proposition 7.3.2.

$$\begin{aligned}
&\left\langle \tilde{M} \right\rangle_t \\
&= \Bigg\langle \int_0^{\cdot} \sum_{i,j,k} \partial_k M^i(ds, x_s) \frac{\partial_j \phi_s^k(x) \, \partial_j \phi_s^i(x)}{\|D\phi_s(x)\|^2} - \partial_k M^i(ds, y_s) \frac{\partial_j \phi_s^k(y) \, \partial_j \phi_s^i(y)}{\|D\phi_s(y)\|^2}, \\
&\quad \int_0^t \sum_{l,m,n} \partial_n M^l(ds, x_s) \frac{\partial_m \phi_s^n(x) \, \partial_m \phi_s^l(x)}{\|D\phi_s(x)\|^2} - \partial_n M^l(ds, y_s) \frac{\partial_m \phi_s^n(y) \, \partial_m \phi_s^l(y)}{\|D\phi_s(y)\|^2} \Bigg\rangle_t \\
&= - \sum_{i,j,k,l,m,n} \int_0^t \frac{\partial_j \phi_s^k(x) \, \partial_j \phi_s^i(x) \, \partial_m \phi_s^n(x) \, \partial_m \phi_s^l(x)}{\|D\phi_s(x)\|^4} \partial_k \partial_n b^{i,l}(0) \, ds \\
&\quad + \sum_{i,j,k,l,m,n} \int_0^t \frac{\partial_j \phi_s^k(x) \, \partial_j \phi_s^i(x) \, \partial_m \phi_s^n(y) \, \partial_m \phi_s^l(y)}{\|D\phi_s(x)\|^2 \|D\phi_s(y)\|^2} \partial_k \partial_n b^{i,l}(x_s - y_s) \, ds \\
&\quad + \sum_{i,j,k,l,m,n} \int_0^t \frac{\partial_j \phi_s^k(y) \, \partial_j \phi_s^i(y) \, \partial_m \phi_s^n(x) \, \partial_m \phi_s^l(x)}{\|D\phi_s(y)\|^2 \|D\phi_s(x)\|^2} \partial_k \partial_n b^{i,l}(y_s - x_s) \, ds \\
&\quad - \sum_{i,j,k,l,m,n} \int_0^t \frac{\partial_j \phi_s^k(y) \, \partial_j \phi_s^i(y) \, \partial_m \phi_s^n(y) \, \partial_m \phi_s^l(y)}{\|D\phi_s(y)\|^4} \partial_k \partial_n b^{i,l}(0) \, ds \\
&= - \sum_{i,j,k,l,m,n} \int_0^t \left[\frac{1}{2}(\beta_N - \beta_L)(\delta_{ki}\delta_{nl} + \delta_{kl}\delta_{ni}) - \beta_N \delta_{kn}\delta_{il} \right] \\
&\quad \left(\frac{\partial_j \phi_s^k(x) \, \partial_j \phi_s^i(x) \, \partial_m \phi_s^n(x) \, \partial_m \phi_s^l(x)}{\|D\phi_s(x)\|^4} + \frac{\partial_j \phi_s^k(y) \, \partial_j \phi_s^i(y) \, \partial_m \phi_s^n(y) \, \partial_m \phi_s^l(y)}{\|D\phi_s(y)\|^4} \right) ds \\
&\quad + 2 \sum_{i,j,k,l,m,n} \int_0^t \frac{\partial_j \phi_s^k(x) \, \partial_j \phi_s^i(x) \, \partial_m \phi_s^n(y) \, \partial_m \phi_s^l(y)}{\|D\phi_s(x)\|^2 \|D\phi_s(y)\|^2} \partial_k \partial_n b^{i,l}(x_s - y_s) \, ds \\
&=: II + III. \quad\quad\quad\quad\quad\quad\quad\quad\quad\quad\quad\quad\quad\quad\quad\quad\quad\quad\quad\quad\quad (7.6)
\end{aligned}$$

Using

$$II = \frac{\beta_L - \beta_N}{2}\int_0^t \sum_{i,j,l,m} \frac{\partial_j\phi_s^i(x)\partial_j\phi_s^i(x)\partial_m\phi_s^l(x)\partial_m\phi_s^l(x)}{\|D\phi_s(x)\|^4}$$
$$+ \frac{\partial_j\phi_s^i(y)\partial_j\phi_s^i(y)\partial_m\phi_s^l(y)\partial_m\phi_s^l(y)}{\|D\phi_s(y)\|^4}ds$$
$$+ \frac{\beta_L - \beta_N}{2}\int_0^t \sum_{i,j,k,m} \frac{\partial_j\phi_s^k(x)\partial_j\phi_s^i(x)\partial_m\phi_s^i(x)\partial_m\phi_s^k(x)}{\|D\phi_s(x)\|^4}$$
$$+ \frac{\partial_j\phi_s^k(y)\partial_j\phi_s^i(y)\partial_m\phi_s^i(y)\partial_m\phi_s^k(y)}{\|D\phi_s(y)\|^4}ds$$
$$+ \beta_N\int_0^t \sum_{i,j,k,m} \frac{\partial_j\phi_s^k(x)\partial_j\phi_s^i(x)\partial_m\phi_s^k(x)\partial_m\phi_s^i(x)}{\|D\phi_s(x)\|^4}$$
$$+ \frac{\partial_j\phi_s^k(y)\partial_j\phi_s^i(y)\partial_m\phi_s^k(y)\partial_m\phi_s^i(y)}{\|D\phi_s(y)\|^4}ds$$
$$= (\beta_L - \beta_N)t + \frac{\beta_L + \beta_N}{2}\int_0^t \sum_{i,j,k,m} \frac{\partial_j\phi_s^k(x)\partial_j\phi_s^i(x)\partial_m\phi_s^k(x)\partial_m\phi_s^i(x)}{\|D\phi_s(x)\|^4}$$
$$+ \frac{\partial_j\phi_s^k(y)\partial_j\phi_s^i(y)\partial_m\phi_s^k(y)\partial_m\phi_s^i(y)}{\|D\phi_s(y)\|^4}ds$$

Proposition 7.3.2 now yields $II \leq (\beta_L - \beta_N)t + \frac{(\beta_N + \beta_L)}{2}2t = 2\beta_L t$.
Since $III \leq 2d^6 \max_{i,l,k,n} \sup_{z \in \mathbb{R}^d} \partial_k\partial_n b^{i,l}(z)\, t$ is clear 3. follows. For the proof of 2. we only have to rearrange (7.6) using the symmetry and isotropy of b.

$$\langle \tilde{M} \rangle_t = -\sum_{i,j,k,l,m,n}\int_0^t \frac{\partial_j\phi_s^k(x)\partial_j\phi_s^i(x)\partial_m\phi_s^n(x)\partial_m\phi_s^l(x)}{\|D\phi_s(x)\|^4}\partial_k\partial_n b^{i,l}(0)\,ds$$
$$+ \sum_{i,j,k,l,m,n}\int_0^t \frac{\partial_j\phi_s^k(x)\partial_j\phi_s^i(x)\partial_m\phi_s^n(y)\partial_m\phi_s^l(y)}{\|D\phi_s(x)\|^2\|D\phi_s(y)\|^2}\partial_k\partial_n b^{i,l}(x_s - y_s)\,ds$$
$$+ \sum_{i,j,k,l,m,n}\int_0^t \frac{\partial_j\phi_s^k(y)\partial_j\phi_s^i(y)\partial_m\phi_s^n(x)\partial_m\phi_s^l(x)}{\|D\phi_s(y)\|^2\|D\phi_s(x)\|^2}\partial_k\partial_n b^{i,l}(y_s - x_s)\,ds$$
$$- \sum_{i,j,k,l,m,n}\int_0^t \frac{\partial_j\phi_s^k(y)\partial_j\phi_s^i(y)\partial_m\phi_s^n(y)\partial_m\phi_s^l(y)}{\|D\phi_s(y)\|^4}\partial_k\partial_n b^{i,l}(0)\,ds$$

7.4 Proof of Lemma 7.2.3: The Two-Point Condition

$$= -\sum_{i,j,k,l,m,n} \int_0^t \left[\frac{\partial_j \phi_s^i(x) \partial_j \phi_s^k(x)}{\|D\phi_s(x)\|^2} - \frac{\partial_j \phi_s^i(y) \partial_j \phi_s^k(y)}{\|D\phi_s(y)\|^2}\right]$$

$$\left[\frac{\partial_m \phi_s^n(x) \partial_m \phi_s^l(x)}{\|D\phi_s(x)\|^2} - \frac{\partial_m \phi_s^n(y) \partial_m \phi_s^l(y)}{\|D\phi_s(y)\|^2}\right] \partial_k \partial_n b^{i,l}(0)\, ds$$

$$+ 2 \sum_{i,j,k,l,m,n} \int_0^t \frac{\partial_j \phi_s^i(x) \partial_j \phi_s^k(x) \partial_m \phi_s^l(y) \partial_m \phi_s^n(y)}{\|D\phi_s(x)\|^2 \|D\phi_s(y)\|^2} \left[\partial_k \partial_n b^{i,l}(x_s - y_s) - \partial_k \partial_n b^{i,l}(0)\right] ds$$

$$=: V + VI.$$

Since (by Lemma 1.3.2) we have $\partial_k \partial_n b^{i,l}(0) = \frac{1}{2}(\beta_N - \beta_L)(\delta_{ki}\delta_{nl} + \delta_{kl}\delta_{ni}) - \beta_N \delta_{kn}\delta_{il}$ we get

$$V = \frac{\beta_L - \beta_N}{2} \sum_{i,j,l,m} \left(\frac{(\partial_j \phi_s^i(x))^2}{\|D\phi_s(x)\|^2} - \frac{(\partial_j \phi_s^i(y))^2}{\|D\phi_s(y)\|^2}\right) \left(\frac{(\partial_m \phi_s^l(x))^2}{\|D\phi_s(x)\|^2} - \frac{(\partial_m \phi_s^l(y))^2}{\|D\phi_s(y)\|^2}\right) ds$$

$$+ \frac{\beta_L - \beta_N}{2} \sum_{i,j,k,m} \int_0^t \left(\frac{\partial_j \phi_s^i(x) \partial_j \phi_s^k(x)}{\|D\phi_s(x)\|^2} - \frac{\partial_j \phi_s^i(y) \partial_j \phi_s^k(y)}{\|D\phi_s(y)\|^2}\right)$$

$$\left(\frac{\partial_m \phi_s^i(x) \partial_m \phi_s^k(x)}{\|D\phi_s(x)\|^2} - \frac{\partial_m \phi_s^i(y) \partial_m \phi_s^k(y)}{\|D\phi_s(y)\|^2}\right) ds$$

$$+ \beta_N \sum_{i,j,k,m} \int_0^t \left(\frac{\partial_j \phi_s^i(x) \partial_j \phi_s^k(x)}{\|D\phi_s(x)\|^2} - \frac{\partial_j \phi_s^i(y) \partial_j \phi_s^k(y)}{\|D\phi_s(y)\|^2}\right)$$

$$\left(\frac{\partial_m \phi_s^k(x) \partial_m \phi_s^i(x)}{\|D\phi_s(x)\|^2} - \frac{\partial_m \phi_s^k(y) \partial_m \phi_s^i(y)}{\|D\phi_s(y)\|^2}\right) ds$$

$$= \frac{\beta_L + \beta_N}{2} \sum_{i,j,k,m} \int_0^t \left(\frac{\partial_j \phi_s^i(x) \partial_j \phi_s^k(x)}{\|D\phi_s(x)\|^2} - \frac{\partial_j \phi_s^i(y) \partial_j \phi_s^k(y)}{\|D\phi_s(y)\|^2}\right)$$

$$\left(\frac{\partial_m \phi_s^i(x) \partial_m \phi_s^k(x)}{\|D\phi_s(x)\|^2} - \frac{\partial_m \phi_s^i(y) \partial_m \phi_s^k(y)}{\|D\phi_s(y)\|^2}\right) ds$$

$$= \frac{\beta_L + \beta_N}{2} \int_0^t \sum_{i,k} \left(\sum_j \frac{\partial_j \phi_s^i(x) \partial_j \phi_s^k(x)}{\|D\phi_s(x)\|^2} - \frac{\partial_j \phi_s^i(y) \partial_j \phi_s^k(y)}{\|D\phi_s(y)\|^2}\right)^2 ds.$$

This completes the proof of Lemma 7.4.1. □

This Lemma shows that it is necessary to control terms of the form $\frac{\partial_j \phi_s^i(x)}{\|D\phi_s(x)\|} - \frac{\partial_j \phi_s^i(y)}{\|D\phi_s(y)\|}$. We postpone this until we will have derived the following estimate from Lemma 7.4.1.

Lemma 7.4.2 (a priori bounds for the ψ-estimation).

1. We have for each $x, y \in \mathbb{R}^d$, $T > 0$ and $q \geq 1$ that

$$\mathbb{E}\left[\sup_{0 \leq t \leq T} |\psi_t(x) - \psi_t(y)|^q\right]^{1/q} \leq (\beta_N + \beta_L)T + 2e^{-\frac{5}{12}}\sqrt{2\pi\tilde{c}}\sqrt{q+1}\sqrt{T}.$$

2. We have for each $x, y \in \mathbb{R}^d$, $T > 0$ and $q \geq 1$ that

$$\mathbb{E}\left[\sup_{0 \leq t \leq T} |\psi_t(x) - \psi_t(y)|^q\right]^{1/q} \leq \bar{C}_1 e^{(\Lambda_1 + \frac{1}{2}\sigma_1^2 q)T}$$

with $\bar{C}_1 := (\beta_N + \beta_L) + 2e^{\frac{1}{4e} - \frac{5}{12}}\sqrt{2\pi\tilde{c}}$, $\Lambda_1 := \frac{1}{e}$ and $\sigma_1 := \sqrt{\frac{2}{e}}$.

3. Let $r > 0$ be fixed. We have for any $x, y \in \mathbb{R}^d$ with $|x - y| \geq r$, $T > 0$ and $q \geq 1$ that

$$\mathbb{E}\left[\sup_{0 \leq t \leq T} |\psi_t(x) - \psi_t(y)|^q\right]^{1/q} \leq \bar{c}_1 |x - y| e^{(\Lambda_1 + \frac{1}{2}\sigma_1^2)T}$$

for $\bar{c}_1 := \frac{1}{r}\left[(\beta_N + \beta_L) + 2e^{-\frac{5}{12}}\sqrt{2\pi\tilde{c}}\right]$ and Λ_1 and σ_1 as before.

Proof: Once again observe that by the triangle inequality

$$\mathbb{E}\left[\sup_{0 \leq t \leq T} |\psi_t(x) - \psi_t(y)|^q\right]^{1/q} \leq \mathbb{E}\left[\sup_{0 \leq t \leq T} A_t^q\right]^{1/q} + \mathbb{E}\left[\sup_{0 \leq t \leq T} \tilde{M}_t^q\right]^{1/q}$$

$$\leq (\beta_N + \beta_L)T + \mathbb{E}\left[\sup_{0 \leq t \leq T} \tilde{M}_t^q\right]^{1/q} =: (\beta_N + \beta_L)T + IV.$$

Since we can write $\tilde{M}_t = W_{\langle \tilde{M} \rangle_t}$ and $\langle \tilde{M} \rangle_t \leq \tilde{c}t$ a.s. we get using [14, (21.24.a)]

$$\mathbb{E}\left[\sup_{0 \leq t \leq T} \tilde{M}_t^q\right] \leq \mathbb{E}\left[\sup_{0 \leq t \leq \tilde{c}T} W_t^q\right] = (\tilde{c}T)^{\frac{q}{2}} \sqrt{\frac{2}{\pi}} \frac{\Gamma\left(\frac{q+1}{2}\right)}{2} 2^{\frac{q+1}{2}}$$

which implies with Lemma 7.1.1

$$IV \leq (2\pi)^{-\frac{1}{2q}} 2^{\frac{q+1}{2q}} \Gamma\left(\frac{q+1}{2}\right)^{\frac{1}{q}} \sqrt{\tilde{c}T} \leq 2e^{-\frac{5}{12}}\sqrt{2\pi\tilde{c}}\sqrt{q+1}\sqrt{T}.$$

7.4 PROOF OF LEMMA 7.2.3: THE TWO-POINT CONDITION

This proves 1.. For the proof of 2. it is sufficient to observe with Lemma 7.1.1

$$(\beta_N + \beta_L)T + 2e^{-\frac{5}{12}}\sqrt{2\pi\tilde{c}}\sqrt{q+1}\sqrt{T} \leq (\beta_N + \beta_L)e^{\frac{T}{e}} + 2e^{-\frac{5}{12}}\sqrt{2\pi\tilde{c}}e^{\frac{\sqrt{q+1}\sqrt{T}}{e}}$$
$$\leq (\beta_N + \beta_L)e^{\frac{T}{e}} + 2e^{-\frac{5}{12}}\sqrt{2\pi\tilde{c}}e^{1/e(1/4+(q+1)T)} \leq \left[(\beta_N + \beta_L) + 2e^{\frac{1}{4e} - \frac{5}{12}}\sqrt{2\pi\tilde{c}}\right]e^{\frac{q+1}{e}T}.$$

This proofs 2. and 3. follows from this by using $\frac{|x-y|}{r} \leq 1$. □

Since we can now control the moments of $\psi_t(x) - \psi_t(y)$ provided x and y are not too close to each other we introduce the following stopping time. Let $\bar{r}$ be chosen according to (1.8) and $\tilde{r} \leq \bar{r}$ to be specified later. Remember that we assumed $\bar{r} \leq 1$. We now define

$$\tau := \inf_{t>0}\{|x_t - y_t| \geq \tilde{r}\}.$$

and will assume $r < \tilde{r}$ in the following (which ensures $\tau > 0$ a.s.). The aim is to estimate $\left(\mathbb{E}\left[\sup_{0 \leq t \leq T} |\psi_t(x) - \psi_t(y)|^q \mathbb{1}_{\{T \leq \tau\}}\right]\right)^{1/q}$ for $|x - y| < r$.

7.4.2 Derivation Of Formula H

We now proceed to work on $\frac{\partial_j \phi_s^i(x)}{\|D\phi_s(x)\|} - \frac{\partial_j \phi_s^i(y)}{\|D\phi_s(y)\|}$ by proving the following proposition on $\frac{\partial_j \phi_s^i(x)}{\|D\phi_s(x)\|}$.

Proposition 7.4.3 (SDE for the direction of the derivative).
We have the SDE

$$\frac{\partial_j \phi_s^i(x)}{\|D\phi_s(x)\|}$$
$$= \frac{\delta^{ij}}{\sqrt{d}} + \int_0^t \sum_k \partial_k M^i(ds, x_s) \frac{\partial_j \phi_s^k(x)}{\|D\phi_s(x)\|} - \int_0^t \sum_{k,l,m} \partial_m M^k(ds, x_s) \frac{\partial_l \phi_s^m(x) \partial_l \phi_s^k(x) \partial_j \phi_s^i(x)}{\|D\phi_s(x)\|^3}$$
$$+ \left[\left(\frac{1}{4} - \frac{d}{2}\right)\beta_N - \frac{1}{4}\beta_L\right]\int_0^t \frac{\partial_j \phi_s^i(x)}{\|D\phi_s(x)\|}ds - \frac{\beta_N + \beta_L}{2}\int_0^t \sum_{k,l} \frac{\partial_j \phi_s^k(x) \partial_l \phi_s^i(x) \partial_l \phi_s^k(x)}{\|D\phi_s(x)\|^3}ds$$
$$+ \frac{3}{4}(\beta_N + \beta_L)\int_0^t \sum_{k,l,m,n} \frac{\partial_l \phi_s^k(x) \partial_l \phi_s^n(x) \partial_m \phi_s^k(x) \partial_m \phi_s^n(x) \partial_j \phi_s^i(x)}{\|D\phi_s(x)\|^5}ds.$$

Proof: Since by Itô's formula

$$\frac{\partial_j \phi_s^i(x)}{\|D\phi_s(x)\|} = \frac{\delta^{ij}}{\sqrt{d}} + \int_0^t \frac{d(\partial_j \phi_\cdot^i(x))_s}{\|D\phi_s(x)\|} - \frac{1}{2}\int_0^t \frac{\partial_j \phi_s^i(x)}{\|D\phi_s(x)\|^3} d\|D\phi_\cdot(x)\|_s^2$$
$$- \frac{1}{2}\int_0^t \frac{d\left\langle \partial_j \phi_\cdot^i(x), \|D\phi_\cdot(x)\|^2 \right\rangle_s}{\|D\phi_s(x)\|^3} + \frac{3}{8}\int_0^t \frac{\partial_j \phi_s^i(x)}{\|D\phi_s(x)\|^5} d\left\langle \|D\phi_\cdot(x)\|^2 \right\rangle_s \quad (7.7)$$

we only have to note that by Theorem 1.3.6, Lemma 1.4.2 and 7.3.1

$$\left\langle \partial_j \phi_\cdot^i(x), \|D\phi_\cdot(x)\|^2 \right\rangle_t$$
$$= \left\langle \int_0^\cdot \sum_k \partial_k M^i(ds, x_s) \partial_j \phi_s^k(x), 2\int_0^\cdot \sum_{l,m,n} \partial_l M^n(ds, x_s) \partial_m \phi_s^l(x) \partial_m \phi_s^n(x) \right\rangle_t$$
$$= -2\int_0^t \sum_{k,l,m,n} \partial_j \phi_s^k(x) \partial_m \phi_s^l(x) \partial_m \phi_s^n(x) \left[\frac{1}{2}(\beta_N - \beta_L)(\delta_{ki}\delta_{ln} + \delta_{kn}\delta_{li}) - \beta_N \delta_{kl}\delta_{in}\right] ds$$
$$= (\beta_L - \beta_N)\int_0^t \sum_{l,m} \partial_j \phi_s^i(x) \partial_m \phi_s^l(x) \partial_m \phi_s^l(x) ds$$
$$+ (\beta_L - \beta_N)\int_0^t \sum_{k,m} \partial_j \phi_s^k(x) \partial_m \phi_s^i(x) \partial_m \phi_s^k(x) ds + 2\beta_N \int_0^t \sum_{k,m} \partial_j \phi_s^k(x) \partial_m \phi_s^k(x) \partial_m \phi_s^i(x) ds$$
$$= (\beta_L - \beta_N)\int_0^t \partial_j \phi_s^i(x) \|D\phi_s(x)\|^2 ds + (\beta_L + \beta_N)\int_0^t \sum_{k,m} \partial_j \phi_s^k(x) \partial_m \phi_s^i(x) \partial_m \phi_s^k(x) ds$$

which combined with Theorem 1.3.6, Lemmas 1.4.2 and 7.3.1 and put into (7.7) yields

$$\frac{\partial_j \phi_t^i(x)}{\|D\phi_t(x)\|} = \frac{\delta^{ij}}{\sqrt{d}} + \int_0^t \sum_k \partial_k M^i(ds, x_s) \frac{\partial_j \phi_s^k(x)}{\|D\phi_s(x)\|} - c\int_0^t \frac{\partial_j \phi_s^i(x)}{\|D\phi_s(x)\|} ds$$
$$- \int_0^t \sum_{k,l,m} \partial_m M^k(ds, x_s) \frac{\partial_l \phi_s^m(x) \partial_l \phi_s^k(x) \partial_j \phi_s^i(x)}{\|D\phi_s(x)\|^3} - \frac{(d-1)\beta_N + \beta_L - 2c}{2}\int_0^t \frac{\partial_j \phi_s^i(x)}{\|D\phi_s(x)\|} ds$$
$$+ \frac{\beta_N - \beta_L}{2}\int_0^t \frac{\partial_j \phi_s^i(x)}{\|D\phi_s(x)\|} ds - \frac{\beta_N + \beta_L}{2}\int_0^t \sum_{k,l} \frac{\partial_j \phi_s^k(x) \partial_l \phi_s^i(x) \partial_l \phi_s^k(x)}{\|D\phi_s(x)\|^3} ds$$
$$+ \frac{3}{4}(\beta_L - \beta_N)\int_0^t \frac{\partial_j \phi_s^i(x)}{\|D\phi_s(x)\|} ds$$
$$+ \frac{3}{4}(\beta_N + \beta_L)\int_0^t \sum_{k,l,m,n} \frac{\partial_l \phi_s^k(x) \partial_l \phi_s^n(x) \partial_m \phi_s^k(x) \partial_m \phi_s^n(x) \partial_j \phi_s^i(x)}{\|D\phi_s(x)\|^5} ds.$$

This proves Proposition 7.4.3. □

7.4 Proof Of Lemma 7.2.3: The Two-Point Condition

Of course the latter implies that $\frac{\partial_j \phi_s^i(x)}{\|D\phi_s(x)\|} - \frac{\partial_j \phi_s^i(y)}{\|D\phi_s(y)\|}$ equals

$$\int_0^t \sum_k \left(\partial_k M^i(ds, x_s) \frac{\partial_j \phi_s^k(x)}{\|D\phi_s(x)\|} - \partial_k M^i(ds, y_s) \frac{\partial_j \phi_s^k(y)}{\|D\phi_s(y)\|} \right)$$
$$- \int_0^t \sum_{k,l,m} \left(\partial_m M^k(ds, x_s) \frac{\partial_l \phi_s^m(x) \partial_l \phi_s^k(x) \partial_j \phi_s^i(x)}{\|D\phi_s(x)\|^3} - \partial_m M^k(ds, y_s) \frac{\partial_l \phi_s^m(y) \partial_l \phi_s^k(y) \partial_j \phi_s^i(y)}{\|D\phi_s(y)\|^3} \right)$$
$$+ \left[\left(\frac{1}{4} - \frac{d}{2} \right) \beta_N - \frac{1}{4} \beta_L \right] \int_0^t \left(\frac{\partial_j \phi_s^i(x)}{\|D\phi_s(x)\|} - \frac{\partial_j \phi_s^i(y)}{\|D\phi_s(y)\|} \right) ds$$
$$- \frac{\beta_N + \beta_L}{2} \int_0^t \sum_{k,l} \left(\frac{\partial_j \phi_s^k(x) \partial_l \phi_s^i(x) \partial_l \phi_s^k(x)}{\|D\phi_s(x)\|^3} - \frac{\partial_j \phi_s^k(y) \partial_l \phi_s^i(y) \partial_l \phi_s^k(y)}{\|D\phi_s(y)\|^3} \right) ds$$
$$+ \frac{3}{4}(\beta_N + \beta_L) \int_0^t \sum_{k,l,m,n} \left(\frac{\partial_l \phi_s^k(x) \partial_l \phi_s^n(x) \partial_m \phi_s^k(x) \partial_m \phi_s^n(x) \partial_j \phi_s^i(x)}{\|D\phi_s(x)\|^5} \right.$$
$$\left. - \frac{\partial_l \phi_s^k(y) \partial_l \phi_s^n(y) \partial_m \phi_s^k(y) \partial_m \phi_s^n(y) \partial_j \phi_s^i(y)}{\|D\phi_s(y)\|^5} \right) ds. \quad (7.8)$$

Letting $VII_t := \int_0^t \sum_k \left(\partial_k M^i(ds, x_s) \frac{\partial_j \phi_s^k(x)}{\|D\phi_s(x)\|} - \partial_k M^i(ds, y_s) \frac{\partial_j \phi_s^k(y)}{\|D\phi_s(y)\|} \right)$ and
$VIII_t := \int_0^t \sum_{k,l,m} \left(\partial_m M^k(ds, x_s) \frac{\partial_l \phi_s^m(x) \partial_l \phi_s^k(x) \partial_j \phi_s^i(x)}{\|D\phi_s(x)\|^3} - \partial_m M^k(ds, y_s) \frac{\partial_l \phi_s^m(y) \partial_l \phi_s^k(y) \partial_j \phi_s^i(y)}{\|D\phi_s(y)\|^3} \right)$ we have to compute the cross variations since we want to apply Itô's formula for powers to (7.8).

$$\langle VII \rangle_t$$
$$= \sum_{k,l} \left\langle \int_0^\cdot \partial_k M^i(ds, x_s) \frac{\partial_j \phi_s^k(x)}{\|D\phi_s(x)\|} - \partial_k M^i(ds, y_s) \frac{\partial_j \phi_s^k(y)}{\|D\phi_s(y)\|}, \right.$$
$$\left. \int_0^\cdot \partial_l M^i(ds, x_s) \frac{\partial_j \phi_s^l(x)}{\|D\phi_s(x)\|} - \partial_l M^i(ds, y_s) \frac{\partial_j \phi_s^l(y)}{\|D\phi_s(y)\|} \right\rangle_t$$
$$= - \int_0^t \sum_{k,l} \left(\frac{\partial_j \phi_s^k(x) \partial_j \phi_s^l(x)}{\|D\phi_s(x)\|^2} + \frac{\partial_j \phi_s^k(y) \partial_j \phi_s^l(y)}{\|D\phi_s(y)\|^2} \right) \partial_k \partial_l b^{i,i}(0) \, ds$$
$$+ \int_0^t \sum_{k,l} \left(\frac{\partial_j \phi_s^k(x) \partial_j \phi_s^l(y)}{\|D\phi_s(x)\| \|D\phi_s(y)\|} + \frac{\partial_j \phi_s^k(y) \partial_j \phi_s^l(x)}{\|D\phi_s(y)\| \|D\phi_s(x)\|} \right) \partial_k \partial_l b^{i,i}(x_s - y_s) \, ds$$
$$= - \sum_{k,l} \int_0^t \left(\frac{\partial_j \phi_s^k(x)}{\|D\phi_s(x)\|} - \frac{\partial_j \phi_s^k(y)}{\|D\phi_s(y)\|} \right) \left(\frac{\partial_j \phi_s^l(x)}{\|D\phi_s(x)\|} - \frac{\partial_j \phi_s^l(y)}{\|D\phi_s(y)\|} \right) [(\beta_N - \beta_L)\delta_{ki}\delta_{li} - \beta_N \delta_{kl}] \, ds$$
$$+ 2\int_0^t \sum_{k,l} \frac{\partial_j \phi_s^k(x) \partial_j \phi_s^l(y)}{\|D\phi_s(x)\| \|D\phi_s(y)\|} \left[\partial_k \partial_l b^{i,i}(x_s - y_s) - \partial_k \partial_l b^{i,i}(0) \right] ds$$

$$=(\beta_L - \beta_N)\int_0^t \left(\frac{\partial_j \phi_s^i(x)}{\|D\phi_s(x)\|} - \frac{\partial_j \phi_s^i(y)}{\|D\phi_s(y)\|}\right)^2 ds + \beta_N \int_0^t \sum_k \left(\frac{\partial_j \phi_s^k(x)}{\|D\phi_s(x)\|} - \frac{\partial_j \phi_s^k(y)}{\|D\phi_s(y)\|}\right)^2 ds$$

$$+ 2\int_0^t \sum_{k,l} \frac{\partial_j \phi_s^k(x) \partial_j \phi_s^l(y)}{\|D\phi_s(x)\| \|D\phi_s(y)\|} \left[\partial_k \partial_l b^{i,i}(x_s - y_s) - \partial_k \partial_l b^{i,i}(0)\right] ds \qquad (7.9)$$

and similarly

$$\langle VIII \rangle_t$$

$$= -\sum_{k,l,m,n,p,r} \int_0^t \left(\frac{\partial_l \phi_s^m(x) \partial_l \phi_s^k(x) \partial_j \phi_s^i(x)}{\|D\phi_s(x)\|^3} - \frac{\partial_l \phi_s^m(y) \partial_l \phi_s^k(y) \partial_j \phi_s^i(y)}{\|D\phi_s(y)\|^3}\right)$$

$$\left(\frac{\partial_p \phi_s^r(x) \partial_p \phi_s^n(x) \partial_j \phi_s^i(x)}{\|D\phi_s(x)\|^3} - \frac{\partial_p \phi_s^r(y) \partial_p \phi_s^n(y) \partial_j \phi_s^i(y)}{\|D\phi_s(y)\|^3}\right)$$

$$\left[\frac{\beta_N - \beta_L}{2}(\delta_{rk}\delta_{mn} + \delta_{rn}\delta_{mk}) - \beta_N \delta_{rm}\delta_{kn}\right] ds$$

$$+ 2\sum_{k,l,m,n,p,r} \int_0^t \frac{\partial_l \phi_s^m(x) \partial_l \phi_s^k(x) \partial_j \phi_s^i(x) \partial_p \phi_s^r(y) \partial_p \phi_s^n(y) \partial_j \phi_s^i(y)}{\|D\phi_s(x)\|^3 \|D\phi_s(y)\|^3}$$

$$\left[\partial_r \partial_m b^{k,n}(x_s - y_s) - \partial_r \partial_m b^{k,n}(0)\right] ds$$

$$= \frac{\beta_L - \beta_N}{2} \sum_{k,l,p,n} \int_0^t \left(\frac{\partial_l \phi_s^n(x) \partial_l \phi_s^k(x) \partial_j \phi_s^i(x)}{\|D\phi_s(x)\|^3} - \frac{\partial_l \phi_s^n(y) \partial_l \phi_s^k(y) \partial_j \phi_s^i(y)}{\|D\phi_s(y)\|^3}\right)$$

$$\left(\frac{\partial_p \phi_s^k(x) \partial_p \phi_s^n(x) \partial_j \phi_s^i(x)}{\|D\phi_s(x)\|^3} - \frac{\partial_p \phi_s^k(y) \partial_p \phi_s^n(y) \partial_j \phi_s^i(y)}{\|D\phi_s(y)\|^3}\right) ds$$

$$+ \frac{\beta_L - \beta_N}{2} \sum_{k,l,n,p} \int_0^t \left(\frac{\partial_l \phi_s^k(x) \partial_l \phi_s^k(x) \partial_j \phi_s^i(x)}{\|D\phi_s(x)\|^3} - \frac{\partial_l \phi_s^k(y) \partial_l \phi_s^k(y) \partial_j \phi_s^i(y)}{\|D\phi_s(y)\|^3}\right)$$

$$\left(\frac{\partial_p \phi_s^n(x) \partial_p \phi_s^n(x) \partial_j \phi_s^i(x)}{\|D\phi_s(x)\|^3} - \frac{\partial_p \phi_s^n(y) \partial_p \phi_s^n(y) \partial_j \phi_s^i(y)}{\|D\phi_s(y)\|^3}\right) ds$$

$$+ \beta_N \sum_{k,l,m,p} \int_0^t \left(\frac{\partial_l \phi_s^m(x) \partial_l \phi_s^k(x) \partial_j \phi_s^i(x)}{\|D\phi_s(x)\|^3} - \frac{\partial_l \phi_s^m(y) \partial_l \phi_s^k(y) \partial_j \phi_s^i(y)}{\|D\phi_s(y)\|^3}\right)$$

$$\left(\frac{\partial_p \phi_s^m(x) \partial_p \phi_s^k(x) \partial_j \phi_s^i(x)}{\|D\phi_s(x)\|^3} - \frac{\partial_p \phi_s^m(y) \partial_p \phi_s^k(y) \partial_j \phi_s^i(y)}{\|D\phi_s(y)\|^3}\right) ds$$

$$+ 2\sum_{k,l,m,n,p,r} \int_0^t \frac{\partial_l \phi_s^m(x) \partial_l \phi_s^k(x) \partial_j \phi_s^i(x) \partial_p \phi_s^r(y) \partial_p \phi_s^n(y) \partial_j \phi_s^i(y)}{\|D\phi_s(x)\|^3 \|D\phi_s(y)\|^3}$$

$$\left[\partial_r \partial_m b^{k,n}(x_s - y_s) - \partial_r \partial_m b^{k,n}(0)\right] ds$$

7.4 Proof Of Lemma 7.2.3: The Two-Point Condition

$$= \frac{\beta_L + \beta_N}{2} \int_0^t \sum_{k,n} \left(\sum_l \frac{\partial_l \phi_s^n(x) \partial_l \phi_s^k(x) \partial_j \phi_s^i(x)}{\|D\phi_s(x)\|^3} - \frac{\partial_l \phi_s^n(y) \partial_l \phi_s^k(y) \partial_j \phi_s^i(y)}{\|D\phi_s(y)\|^3} \right)^2 ds$$

$$+ \frac{\beta_L - \beta_N}{2} \int_0^t \left(\frac{\partial_j \phi_s^i(x)}{\|D\phi_s(x)\|} - \frac{\partial_j \phi_s^i(y)}{\|D\phi_s(y)\|} \right)^2 ds$$

$$+ 2 \sum_{k,l,m,n,p,r} \int_0^t \frac{\partial_l \phi_s^m(x) \partial_l \phi_s^k(x) \partial_j \phi_s^i(x) \partial_p \phi_s^r(y) \partial_p \phi_s^n(y) \partial_j \phi_s^i(y)}{\|D\phi_s(x)\|^3 \|D\phi_s(y)\|^3}$$

$$\left[\partial_r \partial_m b^{k,n}(x_s - y_s) - \partial_r \partial_m b^{k,n}(0) \right] ds \qquad (7.10)$$

as well as

$$\langle VII., VIII.\rangle_t$$
$$= \sum_{k,l,m,n} \left\langle \int_0^{\cdot} \partial_k M^i\left(ds, x_s\right) \frac{\partial_j \phi_s^k\left(x\right)}{\|D\phi_s\left(x\right)\|} - \partial_k M^i\left(ds, y_s\right) \frac{\partial_j \phi_s^k\left(y\right)}{\|D\phi_s\left(y\right)\|}, \right.$$
$$\left. \int_0^{\cdot} \left(\partial_n M^l\left(ds, x_s\right) \frac{\partial_m \phi_s^n\left(x\right) \partial_m \phi_s^l\left(x\right) \partial_j \phi_s^i\left(x\right)}{\|D\phi_s\left(x\right)\|^3} - \partial_n M^l\left(ds, y_s\right) \frac{\partial_m \phi_s^n\left(y\right) \partial_m \phi_s^l\left(y\right) \partial_j \phi_s^i\left(y\right)}{\|D\phi_s\left(y\right)\|^3} \right) \right\rangle_t$$
$$= -\int_0^t \sum_{k,l,m,n} \left(\frac{\partial_j \phi_s^k\left(x\right)}{\|D\phi_s\left(x\right)\|} - \frac{\partial_j \phi_s^k\left(y\right)}{\|D\phi_s\left(y\right)\|} \right) \left[\frac{\beta_N - \beta_L}{2} (\delta_{ik}\delta_{nl} + \delta_{in}\delta_{kl}) - \beta_N \delta_{il}\delta_{kn} \right]$$
$$\left(\frac{\partial_m \phi_s^n\left(x\right) \partial_m \phi_s^l\left(x\right) \partial_j \phi_s^i\left(x\right)}{\|D\phi_s\left(x\right)\|^3} - \frac{\partial_m \phi_s^n\left(y\right) \partial_m \phi_s^l\left(y\right) \partial_j \phi_s^i\left(y\right)}{\|D\phi_s\left(y\right)\|^3} \right) ds$$
$$+ \sum_{k,l,m,n} \int_0^t \left(\frac{\partial_j \phi_s^k\left(x\right) \partial_m \phi_s^n\left(y\right) \partial_m \phi_s^l\left(y\right) \partial_j \phi_s^i\left(y\right)}{\|D\phi_s\left(x\right)\| \|D\phi_s\left(y\right)\|^3} + \frac{\partial_j \phi_s^k\left(y\right) \partial_m \phi_s^n\left(x\right) \partial_m \phi_s^l\left(x\right) \partial_j \phi_s^i\left(x\right)}{\|D\phi_s\left(y\right)\| \|D\phi_s\left(x\right)\|^3} \right)$$
$$\left[\partial_k \partial_n b^{i,l}\left(x_s - y_s\right) - \partial_k \partial_n b^{i,l}\left(0\right) \right] ds$$
$$= \frac{\beta_L - \beta_N}{2} \sum_{l,m} \int_0^t \left(\frac{\partial_j \phi_s^i\left(x\right)}{\|D\phi_s\left(x\right)\|} - \frac{\partial_j \phi_s^i\left(y\right)}{\|D\phi_s\left(y\right)\|} \right)$$
$$\left(\frac{\partial_m \phi_s^l\left(x\right) \partial_m \phi_s^l\left(x\right) \partial_j \phi_s^i\left(x\right)}{\|D\phi_s\left(x\right)\|^3} - \frac{\partial_m \phi_s^l\left(y\right) \partial_m \phi_s^l\left(y\right) \partial_j \phi_s^i\left(y\right)}{\|D\phi_s\left(y\right)\|^3} \right) ds$$
$$+ \frac{\beta_L - \beta_N}{2} \sum_{l,m} \int_0^t \left(\frac{\partial_j \phi_s^l\left(x\right)}{\|D\phi_s\left(x\right)\|} - \frac{\partial_j \phi_s^l\left(y\right)}{\|D\phi_s\left(y\right)\|} \right)$$
$$\left(\frac{\partial_m \phi_s^i\left(x\right) \partial_m \phi_s^l\left(x\right) \partial_j \phi_s^i\left(x\right)}{\|D\phi_s\left(x\right)\|^3} - \frac{\partial_m \phi_s^i\left(y\right) \partial_m \phi_s^l\left(y\right) \partial_j \phi_s^i\left(y\right)}{\|D\phi_s\left(y\right)\|^3} \right) ds$$
$$+ \beta_N \sum_{k,m} \int_0^t \left(\frac{\partial_j \phi_s^k\left(x\right)}{\|D\phi_s\left(x\right)\|} - \frac{\partial_j \phi_s^k\left(y\right)}{\|D\phi_s\left(y\right)\|} \right)$$
$$\left(\frac{\partial_m \phi_s^k\left(x\right) \partial_m \phi_s^i\left(x\right) \partial_j \phi_s^i\left(x\right)}{\|D\phi_s\left(x\right)\|^3} - \frac{\partial_m \phi_s^k\left(y\right) \partial_m \phi_s^i\left(y\right) \partial_j \phi_s^i\left(y\right)}{\|D\phi_s\left(y\right)\|^3} \right) ds$$
$$+ \sum_{k,l,m,n} \int_0^t \left(\frac{\partial_j \phi_s^k\left(x\right) \partial_m \phi_s^n\left(y\right) \partial_m \phi_s^l\left(y\right) \partial_j \phi_s^i\left(y\right)}{\|D\phi_s\left(x\right)\| \|D\phi_s\left(y\right)\|^3} + \frac{\partial_j \phi_s^k\left(y\right) \partial_m \phi_s^n\left(x\right) \partial_m \phi_s^l\left(x\right) \partial_j \phi_s^i\left(x\right)}{\|D\phi_s\left(y\right)\| \|D\phi_s\left(x\right)\|^3} \right)$$
$$\left[\partial_k \partial_n b^{i,l}\left(x_s - y_s\right) - \partial_k \partial_n b^{i,l}\left(0\right) \right] ds$$

$$= \frac{\beta_L - \beta_N}{2} \int_0^t \left(\frac{\partial_j \phi_s^i(x)}{\|D\phi_s(x)\|} - \frac{\partial_j \phi_s^i(y)}{\|D\phi_s(y)\|} \right)^2 ds$$

$$+ \frac{\beta_L + \beta_N}{2} \int_0^t \sum_{k,l} \left(\frac{\partial_j \phi_s^k(x)}{\|D\phi_s(x)\|} - \frac{\partial_j \phi_s^k(y)}{\|D\phi_s(y)\|} \right)$$

$$\left(\frac{\partial_l \phi_s^k(x) \, \partial_l \phi_s^i(x) \, \partial_j \phi_s^i(x)}{\|D\phi_s(x)\|^3} - \frac{\partial_l \phi_s^k(y) \, \partial_l \phi_s^i(y) \, \partial_j \phi_s^i(y)}{\|D\phi_s(y)\|^3} \right) ds$$

$$+ \sum_{k,l,m,n} \int_0^t \left(\frac{\partial_j \phi_s^k(x) \, \partial_m \phi_s^n(y) \, \partial_m \phi_s^l(y) \, \partial_j \phi_s^i(y)}{\|D\phi_s(x)\| \|D\phi_s(y)\|^3} + \frac{\partial_j \phi_s^k(y) \, \partial_m \phi_s^n(x) \, \partial_m \phi_s^l(x) \, \partial_j \phi_s^i(x)}{\|D\phi_s(y)\| \|D\phi_s(x)\|^3} \right)$$

$$\left[\partial_k \partial_n b^{i,l}(x_s - y_s) - \partial_k \partial_n b^{i,l}(0) \right] ds. \tag{7.11}$$

The combination of (7.9), (7.10) and (7.11) with (7.8) and Itô's formula now yields for

$X_t^{(ij)} := \frac{\partial_j \phi_t^i(x)}{\|D\phi_t(x)\|} - \frac{\partial_j \phi_t^i(y)}{\|D\phi_t(y)\|}$ the following (note that $X_t^{(ij)q}$ means $\left(X_t^{(ij)}\right)^q$).

$$X_t^{(ij)q} = q\int_0^t X_s^{(ij)q-1} \sum_k \left(\partial_k M^i(ds,x_s) \frac{\partial_j \phi_s^k(x)}{\|D\phi_s(x)\|} - \partial_k M^i(ds,y_s) \frac{\partial_j \phi_s^k(y)}{\|D\phi_s(y)\|}\right)$$

$$- q\int_0^t X_s^{(ij)q-1} \sum_{k,l,m} \left(\partial_m M^k(ds,x_s) \frac{\partial_l \phi_s^m(x)\partial_l \phi_s^k(x)\partial_j \phi_s^i(x)}{\|D\phi_s(x)\|^3}\right.$$

$$\left. - \partial_m M^k(ds,y_s) \frac{\partial_l \phi_s^m(y)\partial_l \phi_s^k(y)\partial_j \phi_s^i(y)}{\|D\phi_s(y)\|^3}\right) + q\left[\left(\frac{1}{4}-\frac{d}{2}\right)\beta_N - \frac{1}{4}\beta_L\right]\int_0^t X_s^{(ij)q}ds$$

$$- q\frac{\beta_N+\beta_L}{2} \int_0^t X_s^{(ij)q-1} \sum_{k,l} \left(\frac{\partial_j \phi_s^k(x)\partial_l \phi_s^i(x)\partial_l \phi_s^k(x)}{\|D\phi_s(x)\|^3} - \frac{\partial_j \phi_s^k(y)\partial_l \phi_s^i(y)\partial_l \phi_s^k(y)}{\|D\phi_s(y)\|^3}\right) ds$$

$$+ \frac{3}{4}q(\beta_N+\beta_L) \int_0^t X_s^{(ij)q-1} \sum_{k,l,m,n} \left(\frac{\partial_l \phi_s^k(x)\partial_l \phi_s^n(x)\partial_m \phi_s^k(x)\partial_m \phi_s^n(x)\partial_j \phi_s^i(x)}{\|D\phi_s(x)\|^5}\right.$$

$$\left. - \frac{\partial_l \phi_s^k(y)\partial_l \phi_s^n(y)\partial_m \phi_s^k(y)\partial_m \phi_s^n(y)\partial_j \phi_s^i(y)}{\|D\phi_s(y)\|^5}\right) ds$$

$$+ \frac{q(q-1)}{2}(\beta_L-\beta_N)\int_0^t X_s^{(ij)q}ds + \frac{q(q-1)}{2}\beta_N \int_0^t X_s^{(ij)q-2} \sum_k X_s^{(kj)2} ds$$

$$+ q(q-1)\int_0^t X_s^{(ij)q-2} \sum_{k,l} \frac{\partial_j \phi_s^k(x)\partial_j \phi_s^l(y)}{\|D\phi_s(x)\|\|D\phi_s(y)\|} \left[\partial_k\partial_l b^{i,i}(x_s-y_s) - \partial_k\partial_l b^{i,i}(0)\right] ds$$

$$+ q(q-1)\frac{\beta_N+\beta_L}{4}\int_0^t X_s^{(ij)q-2} \sum_{k,n}$$

$$\left(\sum_l \frac{\partial_l \phi_s^n(x)\partial_l \phi_s^k(x)\partial_j \phi_s^i(x)}{\|D\phi_s(x)\|^3} - \frac{\partial_l \phi_s^n(y)\partial_l \phi_s^k(y)\partial_j \phi_s^i(y)}{\|D\phi_s(y)\|^3}\right)^2 ds$$

$$+ q(q-1)\frac{\beta_L-\beta_N}{4}\int_0^t X_s^{(ij)q}ds$$

$$+ q(q-1)\int_0^t X_s^{(ij)q-2} \sum_{k,l,m,n,p,r} \frac{\partial_l \phi_s^m(x)\partial_l \phi_s^k(x)\partial_j \phi_s^i(x)\partial_p \phi_s^r(y)\partial_p \phi_s^n(y)\partial_j \phi_s^i(y)}{\|D\phi_s(x)\|^3 \|D\phi_s(y)\|^3}$$

$$\left[\partial_r\partial_m b^{k,n}(x_s-y_s) - \partial_r\partial_m b^{k,n}(0)\right] ds$$

$$+ q(q-1)\frac{\beta_N-\beta_L}{2}\int_0^t X_s^{(ij)q}ds - q(q-1)\frac{\beta_N+\beta_L}{2}\int_0^t X_s^{(ij)q-2} \sum_{k,l} X_s^{(kj)}$$

$$\left(\frac{\partial_l \phi_s^k(x)\partial_l \phi_s^i(x)\partial_j \phi_s^i(x)}{\|D\phi_s(x)\|^3} - \frac{\partial_l \phi_s^k(y)\partial_l \phi_s^i(y)\partial_j \phi_s^i(y)}{\|D\phi_s(y)\|^3}\right) ds$$

$$- q(q-1)\int_0^t X_s^{(ij)q-2} \sum_{k,l,m,n} \left(\frac{\partial_j \phi_s^k(x)\partial_m \phi_s^n(y)\partial_m \phi_s^l(y)\partial_j \phi_s^i(y)}{\|D\phi_s(x)\|\|D\phi_s(y)\|^3}\right.$$

$$\left. + \frac{\partial_j \phi_s^k(y)\partial_m \phi_s^n(x)\partial_m \phi_s^l(x)\partial_j \phi_s^i(x)}{\|D\phi_s(y)\|\|D\phi_s(x)\|^3}\right) \left[\partial_k\partial_n b^{i,l}(x_s-y_s) - \partial_k\partial_n b^{i,l}(0)\right] ds.$$

7.4 Proof Of Lemma 7.2.3: The Two-Point Condition

Proposition 7.4.4 (Formula H).

$$X_t^{(ij)q} = q \int_0^t X_s^{(ij)q-1} \sum_k \left(\partial_k M^i (ds, x_s) \frac{\partial_j \phi_s^k (x)}{\|D\phi_s (x)\|} - \partial_k M^i (ds, y_s) \frac{\partial_j \phi_s^k (y)}{\|D\phi_s (y)\|} \right)$$

$$- q \int_0^t X_s^{(ij)q-1} \sum_{k,l,m} \left(\partial_m M^k (ds, x_s) \frac{\partial_l \phi_s^m (x) \partial_l \phi_s^k (x) \partial_j \phi_s^i (x)}{\|D\phi_s (x)\|^3} \right.$$

$$\left. - \partial_m M^k (ds, y_s) \frac{\partial_l \phi_s^m (y) \partial_l \phi_s^k (y) \partial_j \phi_s^i (y)}{\|D\phi_s (y)\|^3} \right)$$

$$+ \left[\frac{\beta_L - \beta_N}{4} q^2 - \frac{d-1}{2} \beta_N q - \frac{1}{2} \beta_L q \right] \int_0^t X_s^{(ij)q} ds + \frac{q(q-1)}{2} \beta_N \int_0^t X_s^{(ij)q-2} \sum_k X_s^{(kj)2} ds$$

$$- q \frac{\beta_N + \beta_L}{2} \int_0^t X_s^{(ij)q-1} \sum_{k,l} \left(\frac{\partial_j \phi_s^k (x) \partial_l \phi_s^i (x) \partial_l \phi_s^k (x)}{\|D\phi_s (x)\|^3} - \frac{\partial_j \phi_s^k (y) \partial_l \phi_s^i (y) \partial_l \phi_s^k (y)}{\|D\phi_s (y)\|^3} \right) ds$$

$$+ q(q-1) \frac{\beta_N + \beta_L}{4} \int_0^t X_s^{(ij)q-2} \sum_{k,n}$$

$$\left(\sum_l \frac{\partial_l \phi_s^n (x) \partial_l \phi_s^k (x) \partial_j \phi_s^i (x)}{\|D\phi_s (x)\|^3} - \frac{\partial_l \phi_s^n (y) \partial_l \phi_s^k (y) \partial_j \phi_s^i (y)}{\|D\phi_s (y)\|^3} \right)^2 ds$$

$$- q(q-1) \frac{\beta_N + \beta_L}{2} \int_0^t X_s^{(ij)q-2} \sum_{k,l} X_s^{(kj)}$$

$$\left(\frac{\partial_l \phi_s^k (x) \partial_l \phi_s^i (x) \partial_j \phi_s^i (x)}{\|D\phi_s (x)\|^3} - \frac{\partial_l \phi_s^k (y) \partial_l \phi_s^i (y) \partial_j \phi_s^i (y)}{\|D\phi_s (y)\|^3} \right) ds$$

$$+ \frac{3}{4} q(\beta_N + \beta_L) \int_0^t X_s^{(ij)q-1} \sum_{k,l,m,n} \left(\frac{\partial_l \phi_s^k (x) \partial_l \phi_s^n (x) \partial_m \phi_s^k (x) \partial_m \phi_s^n (x) \partial_j \phi_s^i (x)}{\|D\phi_s (x)\|^5} \right.$$

$$\left. - \frac{\partial_l \phi_s^k (y) \partial_l \phi_s^n (y) \partial_m \phi_s^k (y) \partial_m \phi_s^n (y) \partial_j \phi_s^i (y)}{\|D\phi_s (y)\|^5} \right) ds$$

$$+ q(q-1) \int_0^t X_s^{(ij)q-2} \sum_{k,l} \frac{\partial_j \phi_s^k (x) \partial_j \phi_s^l (y)}{\|D\phi_s (x)\| \|D\phi_s (y)\|} \left[\partial_k \partial_l b^{i,i} (x_s - y_s) - \partial_k \partial_l b^{i,i} (0) \right] ds$$

$$- q(q-1) \int_0^t X_s^{(ij)q-2} \sum_{k,l,m,n} \left(\frac{\partial_j \phi_s^k (x) \partial_m \phi_s^n (y) \partial_m \phi_s^l (y) \partial_j \phi_s^i (y)}{\|D\phi_s (x)\| \|D\phi_s (y)\|^3} \right.$$

$$\left. + \frac{\partial_j \phi_s^k (y) \partial_m \phi_s^n (x) \partial_m \phi_s^l (x) \partial_j \phi_s^i (x)}{\|D\phi_s (y)\| \|D\phi_s (x)\|^3} \right) \left[\partial_k \partial_n b^{i,l} (x_s - y_s) - \partial_k \partial_n b^{i,l} (0) \right] ds$$

$$+ q(q-1) \int_0^t X_s^{(ij)q-2} \sum_{k,l,m,n,p,r} \frac{\partial_l \phi_s^m (x) \partial_l \phi_s^k (x) \partial_j \phi_s^i (x) \partial_p \phi_s^r (y) \partial_p \phi_s^n (y) \partial_j \phi_s^i (y)}{\|D\phi_s (x)\|^3 \|D\phi_s (y)\|^3}$$

$$\left[\partial_r \partial_m b^{k,n} (x_s - y_s) - \partial_r \partial_m b^{k,n} (0) \right] ds. \tag{7.12}$$

Proof: Rearranging terms of the latter equation proofs the proposition. □
Proposition 7.4.4 will be useful to estimate the expectation in Lemma 7.2.3 on the event $\{T \leq \tau\}$ but since this requires some additional preparations we will first consider the reversed case in the following intermezzo.

7.4.3 Treating Small $|x - y|$ And Large T

Since we obviously have from Schwarz' inequality that

$$\mathbb{E}\left[\sup_{0\leq t\leq T} |\psi_t(x) - \psi_t(y)|^q \mathbb{1}_{\{T\geq \tau\}}\right]^{\frac{1}{q}} \leq \mathbb{E}\left[\sup_{0\leq t\leq T} |\psi_t(x) - \psi_t(y)|^{2q}\right]^{\frac{1}{2q}} \mathbb{P}\left[T \geq \tau\right]^{\frac{1}{2q}} \quad (7.13)$$

it seems reasonable to compute some useful estimate for the tails of τ. We will also immediately specify conditions on r and $\tilde{r}$. Assume first with Lemma 1.3.2 that $\tilde{r} < \bar{r}$ is small enough to ensure for any $r \leq \tilde{r}$ that we have

$$\sqrt{2[1 - B_L(r)]} \leq 2\sqrt{\beta_L}r \quad \text{and} \quad \left[(d-1)\frac{1 - B_N(r)}{r} - cr\right] \leq [2(d-1)\beta_N - c]\, r. \quad (7.14)$$

Lemma 7.4.5 (tails of τ).
We have the following estimates for $\mathbb{P}[T \geq \tau]$ and $q \geq \frac{4\sqrt{\beta_L}[2(d-1)\beta_N - c - 2\beta_L]}{128\beta_L} \vee 1$.

1. *If $2(d-1)\beta_N - 2\beta_L - c \leq 0$ and $T \leq \frac{\log\frac{\tilde{r}}{|x-y|}}{16\beta_L q}$ then we have $\mathbb{P}[T \geq \tau] \leq \frac{2}{\tilde{r}^{2q}}|x-y|^{2q}$.*

2. *If $2(d-1)\beta_N - 2\beta_L - c \geq 0$ and $T \leq \frac{\log\frac{\tilde{r}}{|x-y|}}{128\beta_L q} \wedge \frac{\log\frac{\tilde{r}}{|x-y|}}{4\sqrt{\beta_L}[2(d-1)\beta_N - c - 2\beta_L]} = \frac{\log\frac{\tilde{r}}{|x-y|}}{128\beta_L q}$ then we have $\mathbb{P}[T \geq \tau] \leq \frac{2}{\tilde{r}^{2q}}|x-y|^{2q} \wedge \frac{2}{\tilde{r}^{4q}}|x-y|^{4q}$. Of course the latter estimate is also valid in the other case i.e. if $2(d-1)\beta_N - 2\beta_L - c \leq 0$.*

Proof: Let $(X_t)_{t\geq 0}$ be the solution to $X_t = |x-y| + \int_0^t 2\sqrt{\beta_L}X_s dW_s + \int_0^t [2(d-1)\beta_N - c]\, X_s ds$ i.e. let $X_t := |x-y|\exp\left\{2\sqrt{\beta_L}W_s + [2(d-1)\beta_N - c - 2\beta_L]\, t\right\}$ for a BM $(W_t)_{t\geq 0}$. Then we

7.4 Proof of Lemma 7.2.3: The Two-Point Condition

may start with (see (1.12) or (1.21) respectively and (7.14))

$$\begin{aligned}
\mathbb{P}\left[T \geq \tau\right] &= \mathbb{P}\left[\sup_{0 \leq t \leq T} |x_t - y_t| \geq \tilde{r}\right] \leq \mathbb{P}\left[\sup_{0 \leq t \leq T} X_t \geq \tilde{r}\right] \\
&\leq \mathbb{P}\left[\sup_{0 \leq t \leq T} W_t \geq \frac{1}{2\sqrt{\beta_L}} \left(\log \frac{\tilde{r}}{|x-y|} - [2(d-1)\beta_N - 2\beta_L - c]_+ T\right)\right] \\
&= \mathbb{P}\left[\sup_{0 \leq t \leq 1} W_t \geq \frac{\log \frac{\tilde{r}}{|x-y|}}{2\sqrt{\beta_L T}} - [2(d-1)\beta_N - 2\beta_L - c]_+ \sqrt{T}\right] =: IX.
\end{aligned}$$

Let first $2(d-1)\beta_N - 2\beta_L - c \leq 0$. Then we have using Lemma 7.1.1

$$IX \leq 2\left[1 - \Phi\left(\frac{\log \frac{\tilde{r}}{|x-y|}}{2\sqrt{\beta_L T}}\right)\right] \leq 2e^{-\frac{1}{8}\left(\frac{\log \frac{\tilde{r}}{|x-y|}}{\beta_L T}\right)^2} = 2\left(\frac{|x-y|}{\tilde{r}}\right)^{\frac{1}{8\beta_L T} \log \frac{\tilde{r}}{|x-y|}} \leq \frac{2}{\tilde{r}^{2q}} |x-y|^{2q}$$

for $T \leq \frac{\log \frac{\tilde{r}}{|x-y|}}{16\beta_L q}$ (remember $|x-y| < r < \tilde{r}$). Let now $2(d-1)\beta_N - 2\beta_L - c > 0$. In this case we can use for $T \leq \frac{\log \frac{\tilde{r}}{|x-y|}}{4\sqrt{\beta_L}[2(d-1)\beta_N - c - 2\beta_L]} \wedge \frac{\log \frac{\tilde{r}}{|x-y|}}{128\beta_L q}$ that

$$\begin{aligned}
IX &= \mathbb{P}\left[\sup_{0 \leq t \leq 1} W_t \geq \frac{\log \frac{\tilde{r}}{|x-y|}}{2\sqrt{\beta_L T}} - [2(d-1)\beta_N - 2\beta_L - c]\sqrt{T}\right] \\
&\leq \mathbb{P}\left[\sup_{0 \leq t \leq 1} W_t \geq \frac{\log \frac{\tilde{r}}{|x-y|}}{4\sqrt{\beta_L T}}\right] = 2\left[1 - \Phi\left(\frac{\log \frac{\tilde{r}}{|x-y|}}{4\sqrt{\beta_L T}}\right)\right] \\
&\leq 2\exp\left\{-\frac{1}{2}\frac{1}{16\beta_L T}\left(\log \frac{\tilde{r}}{|x-y|}\right)^2\right\} = 2\left(\frac{|x-y|}{\tilde{r}}\right)^{\frac{\log \frac{\tilde{r}}{|x-y|}}{32\beta_L T}} \leq 2\frac{|x-y|^{2q}}{\tilde{r}^{2q}} \wedge 2\frac{|x-y|^{4q}}{\tilde{r}^{4q}}.
\end{aligned}$$

The proof is complete. □

Lemma 7.4.6 (estimate for T after τ).
For $|x-y| \leq r \leq e^{-1}\tilde{r}$ and arbitrary $q \geq \frac{4\sqrt{\beta_L}[2(d-1)\beta_N - c - 2\beta_L]}{128\beta_L} \vee 1$ and $T > 0$ we have the estimate $\mathbb{E}\left[\sup_{0 \leq t \leq T} |\psi_t(x) - \psi_t(y)|^q \mathbb{1}_{\{T \geq \tau\}}\right]^{1/q} \leq \bar{c}_2 |x-y|^2 e^{(\Lambda_2 + \frac{1}{2}q\sigma_2^2)T}$ for $\Lambda_2 := \Lambda_1 \vee 1$, $\sigma_2 := \sqrt{2}\sigma_1 \vee 2\sqrt{128\beta_L} \vee 2$ and $\bar{c}_2 := \frac{\sqrt{2}\bar{C}_1}{\tilde{r}} \vee \frac{1}{\tilde{r}}\left[\frac{\beta_N + \beta_L}{128\beta_L} + \frac{\sqrt{2\pi}\bar{c}e^{-\frac{5}{12}}}{\sqrt{16\beta_L}}\right] \vee \frac{2(\beta_L + \beta_N)}{\tilde{r}} \vee \frac{2e^{-\frac{5}{12}}\sqrt{256\bar{c}\pi\beta_L}}{\tilde{r}}$.

Proof: If $T \leq \frac{\log \frac{\tilde{r}}{|x-y|}}{128\beta_L q} \wedge \frac{\log \frac{\tilde{r}}{|x-y|}}{4\sqrt{\beta_L}[2(d-1)\beta_N - c - 2\beta_L]} = \frac{\log \frac{\tilde{r}}{|x-y|}}{128\beta_L q}$ then we just have to combine

Lemmas 7.4.2 and 7.4.5 with (7.13) to obtain

$$\mathbb{E}\left[\sup_{0\leq t\leq T}|\psi_t(x)-\psi_t(y)|^q \mathbb{1}_{\{T\geq \tau\}}\right]^{1/q} \leq |x-y|\frac{\bar{C}_1}{\tilde{r}}2^{\frac{1}{2q}}e^{(\Lambda_1+\frac{1}{2}\sigma_1^2 2q)T}$$

which means we only have to consider the case $T \geq \frac{\log\frac{\tilde{r}}{|x-y|}}{128\beta_L q}$. By Lemma 7.4.2 we first observe for $T_0 := \frac{\log\frac{\tilde{r}}{|x-y|}}{128\beta_L q}$ that

$$\mathbb{E}\left[\sup_{0\leq t\leq T_0}|\psi_t(x)-\psi_t(y)|^q\right]^{1/q}$$
$$\leq (\beta_N+\beta_L)\frac{\log\frac{\tilde{r}}{|x-y|}}{128\beta_L q}$$
$$+2e^{-\frac{5}{12}}\sqrt{2\pi}\tilde{c}\sqrt{q+1}\sqrt{\frac{\log\frac{\tilde{r}}{|x-y|}}{128\beta_L q}} \leq \left[\frac{\beta_N+\beta_L}{128\beta_L}+\frac{\sqrt{2\pi}\tilde{c}e^{-\frac{5}{12}}}{\sqrt{16\beta_L}}\right]\frac{\tilde{r}}{|x-y|}$$
$$\leq \left[\frac{\beta_N+\beta_L}{128\beta_L}+\frac{\sqrt{2\pi}\tilde{c}e^{-\frac{5}{12}}}{\sqrt{16\beta_L}}\right]\left(\frac{\tilde{r}}{|x-y|}\right)^{\frac{\Lambda_2+\frac{1}{2}\sigma_2^2 q}{128\beta_L q}-1}$$
$$= \frac{1}{\tilde{r}}\left[\frac{\beta_N+\beta_L}{128\beta_L}+\frac{\sqrt{2\pi}\tilde{c}e^{-\frac{5}{12}}}{\sqrt{16\beta_L}}\right]|x-y|e^{\frac{\Lambda_2+\frac{1}{2}\sigma_2^2 q}{128\beta_L q}\log\frac{\tilde{r}}{|x-y|}} \leq \bar{c}_2|x-y|e^{(\Lambda_2+\frac{1}{2}q\sigma_2^2)T_0}.$$

Since $f : T \mapsto \bar{c}_2|x-y|e^{(\Lambda_2+\frac{1}{2}q\sigma_2^2)T}$ is a convex function and $g : T \mapsto (\beta_N+\beta_L)T + 2e^{-\frac{5}{12}}\sqrt{2\pi}\tilde{c}\sqrt{q+1}\sqrt{T}$ is concave one we may just check that $f'(T_0) \geq g'(T_0)$. First let us fix

$$g'(T) = \beta_N+\beta_L + \frac{e^{-\frac{5}{12}}\sqrt{2\pi}\tilde{c}\sqrt{q+1}}{\sqrt{T}} \quad \text{and} \quad f'(T) = \bar{c}_2|x-y|(\Lambda_2+\frac{1}{2}q\sigma_2^2)e^{(\Lambda_2+\frac{1}{2}q\sigma_2^2)T}.$$

So we can compute

$$g'(T_0) = (\beta_N+\beta_L) + e^{-\frac{5}{12}}\sqrt{256\pi\tilde{c}\beta_L}\sqrt{q(q+1)}\frac{1}{\log\frac{\tilde{r}}{|x-y|}}$$
$$\leq \frac{\bar{c}_2\tilde{r}}{2}(\Lambda_2+\frac{1}{2}q\sigma_2^2)\left(\frac{\tilde{r}}{|x-y|}\right)^{\frac{\Lambda_2+\frac{1}{2}\sigma_2^2 q}{128\beta_L}-1} + \frac{\bar{c}_2\tilde{r}}{2}(\Lambda_2+\frac{1}{2}q\sigma_2^2)\left(\frac{\tilde{r}}{|x-y|}\right)^{\frac{\Lambda_2+\frac{1}{2}\sigma_2^2 q}{128\beta_L}-1}$$

7.4 Proof Of Lemma 7.2.3: The Two-Point Condition

$$= \bar{c}_2(\Lambda_2 + \frac{1}{2}q\sigma_2^2)\left(\frac{\tilde{r}}{|x-y|}\right)^{\frac{\Lambda_2 + \frac{1}{2}\sigma_2^2 q}{128\beta_L}} |x-y| = f'(T_0).$$

Thus the proof of Lemma 7.4.6 is complete. □

To treat (7.12) we finally need the following proposition.

Proposition 7.4.7 (expectation of zero-mean-martingales on rare events).
Let for $i, j \in \{1, \ldots, d\}$

$$\check{M}_t = \check{M}_t^{(ij)} := \int_0^t X_s^{(ij)q-1} dVII_t - \int_0^t X_s^{(ij)q-1} dVIII_t.$$

Then we have the following.

1. $\check{M}_t$ is a zero-mean-martingale and we have for any $t \geq 0$ that $\langle \check{M} \rangle_t \leq \check{c} t$ a.s. wherein we define the constant $\check{c} := 4(d^2 + 2d^4 + d^6) \max_{k,l,m,n} \sup_{z \in \mathbb{R}^d} \partial_k \partial_l b^{m,n}(z)$.

2. For all $0 \leq t \leq T$ and $q \geq \frac{4\sqrt{\beta_L}[2(d-1)\beta_N - c - 2\beta_L]}{128\beta_L} \vee 1$ we have

$$\left|\mathbb{E}\left[\check{M}_t \mathbb{1}_{\{T \leq \tau\}}\right]\right| \leq |x-y|^{2q} \frac{\bar{c}_3}{\tilde{r}^{2q}} e^{(\Lambda_3 + \frac{1}{2}\sigma_3^2 q + \sigma_4^2 q^2)T} =: f(T)$$

for $\bar{c}_3 := \sqrt{2\check{c}} \vee \sqrt{\frac{\check{c}}{128\beta_L}} \vee 1$, $\Lambda_3 := \frac{1}{2e}$, $\sigma_3 := \sqrt{\frac{64\beta_L}{e}} \vee (128\beta_L \check{c})^{\frac{1}{4}}$ and $\sigma_4 := \sqrt{256\beta_L}$.

Proof: 1. is clear from the definition of $\check{M}$ except for the value of $\check{c}$. This value follows from (7.9), (7.10) and (7.11) since $\langle \check{M} \rangle_t \leq \langle VII \rangle_i + \langle VIII \rangle_t + 2|\langle VII, VIII \rangle_t|$. For the proof of 2. first assume $T \leq \frac{\log \frac{\tilde{r}}{|x-y|}}{128\beta_L q}$. From Lemma 7.4.5 we know that $\mathbb{P}[T > \tau] \leq |x-y|^{4q} \frac{2}{\tilde{r}^{4q}}$. This combined with Schwarz' inequality and 1. yields

$$\left|\mathbb{E}\left[\check{M}_t \mathbb{1}_{\{T \leq \tau\}}\right]\right| = \left|\mathbb{E}\left[\check{M}_t \mathbb{1}_{\{T > \tau\}}\right]\right| \leq \sqrt{\mathbb{E}[\langle \check{M} \rangle_T]} \sqrt{\mathbb{P}[T > \tau]} \leq \sqrt{2\check{c}T} \frac{|x-y|^{2q}}{\tilde{r}^{2q}} \leq \sqrt{2\check{c}} \frac{|x-y|^{2q}}{\tilde{r}^{2q}} e^{\frac{T}{2e}}$$

which proves Proposition 7.4.7 in this case. We always have as above that

$\left|\mathbb{E}\left[\check{M}_t \mathbb{1}_{\{T \leq \tau\}}\right]\right| \leq \sqrt{\check{c}T} =: g(T)$ and so we can conclude for the same T_0 as before

$$g(T_0) = \sqrt{\frac{\check{c}}{128\beta_L}} \sqrt{\log \frac{\tilde{r}}{|x-y|}} \leq \sqrt{\frac{\check{c}}{128\beta_L}} \left(\frac{\tilde{r}}{|x-y|}\right)^{\frac{1}{2e}} \leq \bar{c}_3 \left(\frac{\tilde{r}}{|x-y|}\right)^{\frac{\sigma_3^2}{128\beta_L}}$$

$$\leq \bar{c}_3 \left(\frac{\tilde{r}}{|x-y|}\right)^{\frac{\Lambda_3 + \sigma_3^2 q + \sigma_4^2 q^2}{128\beta_L q}} \left(\frac{|x-y|}{\tilde{r}}\right)^{2q} = |x-y|^{2q} \frac{\bar{C}_3}{\tilde{r}^{2q}} e^{(\Lambda_3 + \frac{1}{2}\sigma_3^2 q + \sigma_4^2 q^2)T_0} =: f(T_0).$$

So now (with Lemma 7.1.2) we only have to establish the inequality $g'(T_0) \leq f'(T_0)$ to complete the proof of Proposition 7.4.7.

$$g'(T_0) = \frac{1}{2} \sqrt{\frac{128\beta_L q \check{c}}{\log \frac{\tilde{r}}{|x-y|}}}$$

$$\leq \frac{1}{2} \sqrt{128\beta_L q \check{c}} \leq \left(\frac{|x-y|}{\tilde{r}}\right)^{2q} \left(\Lambda_3 + \frac{1}{2}\sigma_3^2 q + \sigma_4^2 q^2\right) \left(\frac{\tilde{r}}{|x-y|}\right)^{\frac{\Lambda_3 + \frac{1}{2}\sigma_3^2 q + \sigma_4^2 q^2}{128\beta_L}} = f'(T_0).$$

The proof is complete. $\square$

7.4.4 Evaluation Of Formula H

Now we are prepared to prove the following key result

Lemma 7.4.8 (first estimate for τ after T).
We have for even $q \geq q_0$ if we assume $\tilde{r} \leq \frac{1}{\sqrt{2}}$ the following.

1. *For $h(t) := \mathbb{E}\left[(\max_{i,j} X_t^{(ij)q} \vee |x_t - y_t|^q)\mathbb{1}_{\{t \leq \tau\}}\right]$ the estimate*

$$h(t) \leq |x-y|^q \frac{2\bar{c}_3 d^2 q}{\tilde{r}^{2q}} \frac{e^{(\Lambda_3 + (\lambda \vee \frac{1}{2}\sigma_3^2)q + (\frac{1}{2}\bar{\sigma}^2 \vee \sigma_4^2)q^2 + d^2\kappa_q)t}}{\Lambda_3 + (\lambda \vee \frac{1}{2}\sigma_3^2)q + (\frac{1}{2}\bar{\sigma}^2 \vee \sigma_4^2)q^2 + d^2\kappa_q}$$

holds wherein κ_q behaves like a polynomial of degree 2 (w.r.t. growths of its modulus) in q and equals $\left|\frac{\beta_L - \beta_N}{4}q^2 - \frac{d-1}{2}\beta_N q - \frac{1}{2}\beta_L q\right|$
$+ \left\{Cd^6 + \left[\frac{9}{4}(\beta_N + \beta_L) + 2C\right]d^4 + \left[\frac{3}{2}(\beta_N + \beta_L) + C\right]d^2 + \frac{\beta_N}{2}d\right\}q^2$
$- \left\{Cd^6 + \left[2C - \frac{3}{2}(\beta_N + \beta_L)\right]d^4 + Cd^2 + \frac{1}{2}\beta_N d\right\}q.$

2. *We have*

$$\mathbb{E}\left[\max_{i,j} X_t^{(ij)q} \mathbb{1}_{\{T \leq \tau\}}\right] \leq h(t) \leq \bar{c}_5 \tilde{r}^{-2q}|x-y|^q e^{(\Lambda_5 q + \sigma_5^2 q^2)t}$$

for $\bar{c}_5 := \sup_{q \geq 3}\left\{\frac{2\bar{c}_3 d^2 q}{\Lambda_3 + (\lambda \vee \frac{1}{2}\sigma_3^2)q + (\frac{1}{2}\bar{\sigma}^2 \vee \sigma_4^2)q^2 + d^2\kappa_q}\right\},$
$\Lambda_5 := \left(\Lambda_3 + \lambda \vee \frac{1}{2}\sigma_3^2 - \left\{Cd^8 + \left[2C - \frac{3}{2}(\beta_N + \beta_L)\right]d^6 + Cd^4 - \frac{|\beta_N - \beta_L|}{2}d^2\right\}\right)_+$ and
$\sigma_5^2 := \frac{1}{2}\bar{\sigma}^2 \vee \sigma_4^2$
$+ \left\{Cd^8 + \left[\frac{9}{4}(\beta_N + \beta_L) + 2C\right]d^6 + \left[\frac{3}{2}(\beta_N + \beta_L) + C\right]d^4 + \frac{\beta_N}{2}d^3 + \frac{|\beta_L - \beta_N|}{4}d^2\right\}.$

Proof: 2. follows obviously from 1. so we only have to prove this. First observe that we have by Proposition 7.4.4 that

$$X_t^{(ij)q}\mathbb{1}_{\{t\le\tau\}}$$
$$=q\mathbb{1}_{\{t\le\tau\}}\int_0^t X_s^{(ij)q-1}\sum_k\left(\partial_k M^i(ds,x_s)\frac{\partial_j\phi_s^k(x)}{\|D\phi_s(x)\|}-\partial_k M^i(ds,y_s)\frac{\partial_j\phi_s^k(y)}{\|D\phi_s(y)\|}\right)$$
$$-q\mathbb{1}_{\{t\le\tau\}}\int_0^t X_s^{(ij)q-1}\sum_{k,l,m}\left(\partial_m M^k(ds,x_s)\frac{\partial_l\phi_s^m(x)\,\partial_l\phi_s^k(x)\,\partial_j\phi_s^i(x)}{\|D\phi_s(x)\|^3}\right.$$
$$\left.-\partial_m M^k(ds,y_s)\frac{\partial_l\phi_s^m(y)\,\partial_l\phi_s^k(y)\,\partial_j\phi_s^i(y)}{\|D\phi_s(y)\|^3}\right)$$
$$+\left[\frac{\beta_L-\beta_N}{4}q^2-\frac{d-1}{2}\beta_N q-\frac{1}{2}\beta_L q\right]\int_0^t X_s^{(ij)q}\mathbb{1}_{\{t\le\tau\}}ds$$
$$+\frac{q(q-1)}{2}\beta_N\int_0^t X_s^{(ij)q-2}\sum_k X_s^{(kj)2}\mathbb{1}_{\{t\le\tau\}}ds$$
$$-q\frac{\beta_N+\beta_L}{2}\int_0^t X_s^{(ij)q-1}\sum_{k,l}\left(\frac{\partial_j\phi_s^k(x)\,\partial_l\phi_s^i(x)\,\partial_l\phi_s^k(x)}{\|D\phi_s(x)\|^3}-\frac{\partial_j\phi_s^k(y)\,\partial_l\phi_s^i(y)\,\partial_l\phi_s^k(y)}{\|D\phi_s(y)\|^3}\right)\mathbb{1}_{\{t\le\tau\}}ds$$
$$+q(q-1)\frac{\beta_N+\beta_L}{4}\int_0^t X_s^{(ij)q-2}\sum_{k,n}\left(\sum_l\frac{\partial_l\phi_s^n(x)\,\partial_l\phi_s^k(x)\,\partial_j\phi_s^i(x)}{\|D\phi_s(x)\|^3}\right.$$
$$\left.-\frac{\partial_l\phi_s^n(y)\,\partial_l\phi_s^k(y)\,\partial_j\phi_s^i(y)}{\|D\phi_s(y)\|^3}\right)^2\mathbb{1}_{\{t\le\tau\}}ds$$
$$-q(q-1)\frac{\beta_N+\beta_L}{2}\int_0^t X_s^{(ij)q-2}\sum_{k,l}X_s^{(kj)}\left(\frac{\partial_l\phi_s^k(x)\,\partial_l\phi_s^i(x)\,\partial_j\phi_s^i(x)}{\|D\phi_s(x)\|^3}\right.$$
$$\left.-\frac{\partial_l\phi_s^k(y)\,\partial_l\phi_s^i(y)\,\partial_j\phi_s^i(y)}{\|D\phi_s(y)\|^3}\right)\mathbb{1}_{\{t\le\tau\}}ds$$
$$+\frac{3}{4}q(\beta_N+\beta_L)\int_0^t X_s^{(ij)q-1}\sum_{k,l,m,n}\left(\frac{\partial_l\phi_s^k(x)\,\partial_l\phi_s^n(x)\,\partial_m\phi_s^k(x)\,\partial_m\phi_s^n(x)\,\partial_j\phi_s^i(x)}{\|D\phi_s(x)\|^5}\right.$$
$$\left.-\frac{\partial_l\phi_s^k(y)\,\partial_l\phi_s^n(y)\,\partial_m\phi_s^k(y)\,\partial_m\phi_s^n(y)\,\partial_j\phi_s^i(y)}{\|D\phi_s(y)\|^5}\right)\mathbb{1}_{\{t\le\tau\}}ds$$
$$+q(q-1)\int_0^t X_s^{(ij)q-2}\sum_{k,l}\frac{\partial_j\phi_s^k(x)\,\partial_j\phi_s^l(y)}{\|D\phi_s(x)\|\,\|D\phi_s(y)\|}\left[\partial_k\partial_l b^{i,i}(x_s-y_s)-\partial_k\partial_l b^{i,i}(0)\right]\mathbb{1}_{\{t\le\tau\}}ds$$
$$-q(q-1)\int_0^t X_s^{(ij)q-2}\sum_{k,l,m,n}\left(\frac{\partial_j\phi_s^k(x)\,\partial_m\phi_s^n(y)\,\partial_m\phi_s^l(y)\,\partial_j\phi_s^i(y)}{\|D\phi_s(x)\|\,\|D\phi_s(y)\|^3}\right.$$
$$\left.+\frac{\partial_j\phi_s^k(y)\,\partial_m\phi_s^n(x)\,\partial_m\phi_s^l(x)\,\partial_j\phi_s^i(x)}{\|D\phi_s(y)\|\,\|D\phi_s(x)\|^3}\right)\left[\partial_k\partial_n b^{i,l}(x_s-y_s)-\partial_k\partial_n b^{i,l}(0)\right]\mathbb{1}_{\{t\le\tau\}}ds$$
$$+q(q-1)\int_0^t X_s^{(ij)q-2}\sum_{k,l,m,n,p,r}\frac{\partial_l\phi_s^m(x)\,\partial_l\phi_s^k(x)\,\partial_j\phi_s^i(x)\,\partial_p\phi_s^r(y)\,\partial_p\phi_s^n(y)\,\partial_j\phi_s^i(y)}{\|D\phi_s(x)\|^3\,\|D\phi_s(y)\|^3}$$
$$\left[\partial_r\partial_m b^{k,n}(x_s-y_s)-\partial_r\partial_m b^{k,n}(0)\right]\mathbb{1}_{\{t\le\tau\}}ds.$$

7.4 Proof Of Lemma 7.2.3: The Two-Point Condition

Taking expectations and applying Fubini's theorem we get

$$\mathbb{E}\left[X_t^{(ij)q}\mathbb{1}_{\{t\leq\tau\}}\right]$$
$$=q\mathbb{E}\left[\mathbb{1}_{\{t\leq\tau\}}\check{M}_t^{(ij)}\right]$$
$$+\left[\frac{\beta_L-\beta_N}{4}q^2-\frac{d-1}{2}\beta_N q-\frac{1}{2}\beta_L q\right]\int_0^t\mathbb{E}\left[X_s^{(ij)q}\mathbb{1}_{\{t\leq\tau\}}\right]ds$$
$$+\frac{q(q-1)}{2}\beta_N\int_0^t\mathbb{E}\left[X_s^{(ij)q-2}\sum_k X_s^{(kj)2}\mathbb{1}_{\{t\leq\tau\}}\right]ds$$
$$-q\frac{\beta_N+\beta_L}{2}\int_0^t\mathbb{E}\left[X_s^{(ij)q-1}\sum_{k,l}\left(\frac{\partial_j\phi_s^k(x)\,\partial_l\phi_s^i(x)\,\partial_l\phi_s^k(x)}{\|D\phi_s(x)\|^3}-\frac{\partial_j\phi_s^k(y)\,\partial_l\phi_s^i(y)\,\partial_l\phi_s^k(y)}{\|D\phi_s(y)\|^3}\right)\right.$$
$$\left.\mathbb{1}_{\{t\leq\tau\}}\right]ds+q(q-1)\frac{\beta_N+\beta_L}{4}\int_0^t\mathbb{E}\left[X_s^{(ij)q-2}\sum_{k,n}\right.$$
$$\left(\sum_l\frac{\partial_l\phi_s^n(x)\,\partial_l\phi_s^k(x)\,\partial_j\phi_s^i(x)}{\|D\phi_s(x)\|^3}-\frac{\partial_l\phi_s^n(y)\,\partial_l\phi_s^k(y)\,\partial_j\phi_s^i(y)}{\|D\phi_s(y)\|^3}\right)^2\mathbb{1}_{\{t\leq\tau\}}\right]ds$$
$$-q(q-1)\frac{\beta_N+\beta_L}{2}\int_0^t\mathbb{E}\left[X_s^{(ij)q-2}\sum_{k,l}\right.$$
$$\left.X_s^{(kj)}\left(\frac{\partial_l\phi_s^k(x)\,\partial_l\phi_s^i(x)\,\partial_j\phi_s^i(x)}{\|D\phi_s(x)\|^3}-\frac{\partial_l\phi_s^k(y)\,\partial_l\phi_s^i(y)\,\partial_j\phi_s^i(y)}{\|D\phi_s(y)\|^3}\right)\mathbb{1}_{\{t\leq\tau\}}\right]ds$$
$$+\frac{3}{4}q(\beta_N+\beta_L)\int_0^t\mathbb{E}\left[X_s^{(ij)q-1}\sum_{k,l,m,n}\left(\frac{\partial_l\phi_s^k(x)\,\partial_l\phi_s^n(x)\,\partial_m\phi_s^k(x)\,\partial_m\phi_s^n(x)\,\partial_j\phi_s^i(x)}{\|D\phi_s(x)\|^5}\right.\right.$$
$$\left.\left.-\frac{\partial_l\phi_s^k(y)\,\partial_l\phi_s^n(y)\,\partial_m\phi_s^k(y)\,\partial_m\phi_s^n(y)\,\partial_j\phi_s^i(y)}{\|D\phi_s(y)\|^5}\right)\mathbb{1}_{\{t\leq\tau\}}\right]ds$$
$$+q(q-1)\int_0^t\mathbb{E}\left[X_s^{(ij)q-2}\sum_{k,l}\frac{\partial_j\phi_s^k(x)\,\partial_j\phi_s^l(y)}{\|D\phi_s(x)\|\,\|D\phi_s(y)\|}\left[\partial_k\partial_l b^{i,i}(x_s-y_s)-\partial_k\partial_l b^{i,i}(0)\right]\mathbb{1}_{\{t\leq\tau\}}\right]ds$$
$$-q(q-1)\int_0^t\mathbb{E}\left[X_s^{(ij)q-2}\sum_{k,l,m,n}\left(\frac{\partial_j\phi_s^k(x)\,\partial_m\phi_s^n(y)\,\partial_m\phi_s^l(y)\,\partial_j\phi_s^i(y)}{\|D\phi_s(x)\|\,\|D\phi_s(y)\|^3}\right.\right.$$
$$\left.\left.+\frac{\partial_j\phi_s^k(y)\,\partial_m\phi_s^n(x)\,\partial_m\phi_s^l(x)\,\partial_j\phi_s^i(x)}{\|D\phi_s(y)\|\,\|D\phi_s(x)\|^3}\right)\left[\partial_k\partial_n b^{i,l}(x_s-y_s)-\partial_k\partial_n b^{i,l}(0)\right]\mathbb{1}_{\{t\leq\tau\}}\right]ds$$
$$+q(q-1)\int_0^t\mathbb{E}\left[X_s^{(ij)q-2}\sum_{k,l,m,n,p,r}\frac{\partial_l\phi_s^m(x)\,\partial_l\phi_s^k(x)\,\partial_j\phi_s^i(x)\,\partial_p\phi_s^r(y)\,\partial_p\phi_s^n(y)\,\partial_j\phi_s^i(y)}{\|D\phi_s(x)\|^3\,\|D\phi_s(y)\|^3}\right.$$
$$\left.\left[\partial_r\partial_m b^{k,n}(x_s-y_s)-\partial_r\partial_m b^{k,n}(0)\right]\mathbb{1}_{\{t\leq\tau\}}\right]ds.$$

Inserting some moduli and using Lemma 1.3.2 we get

$$\mathbb{E}\left[X_t^{(ij)q}\mathbb{1}_{\{t\leq\tau\}}\right]$$
$$\leq q\mathbb{E}\left[\mathbb{1}_{\{t\leq\tau\}}\check{M}_t^{(ij)}\right]$$
$$+\left|\frac{\beta_L-\beta_N}{4}q^2-\frac{d-1}{2}\beta_N q-\frac{1}{2}\beta_L q\right|\int_0^t \mathbb{E}\left[X_s^{(ij)q}\mathbb{1}_{\{t\leq\tau\}}\right]ds$$
$$+\frac{q(q-1)}{2}\beta_N\int_0^t \mathbb{E}\left[X_s^{(ij)q-2}\sum_k X_s^{(kj)2}\mathbb{1}_{\{t\leq\tau\}}\right]ds$$
$$+q\frac{\beta_N+\beta_L}{2}\int_0^t \mathbb{E}\Bigg[\left|X_s^{(ij)q-1}\right|\sum_{k,l}\left|\frac{\partial_j\phi_s^k(x)\partial_l\phi_s^i(x)\partial_l\phi_s^k(x)}{\|D\phi_s(x)\|^3}-\frac{\partial_j\phi_s^k(y)\partial_l\phi_s^i(y)\partial_l\phi_s^k(y)}{\|D\phi_s(y)\|^3}\right|$$
$$\mathbb{1}_{\{t\leq\tau\}}\Bigg]ds+q(q-1)\frac{\beta_N+\beta_L}{4}\int_0^t \mathbb{E}\Bigg[X_s^{(ij)q-2}\sum_{k,n}$$
$$\left(\sum_l\frac{\partial_l\phi_s^n(x)\partial_l\phi_s^k(x)\partial_j\phi_s^i(x)}{\|D\phi_s(x)\|^3}-\frac{\partial_l\phi_s^n(y)\partial_l\phi_s^k(y)\partial_j\phi_s^i(y)}{\|D\phi_s(y)\|^3}\right)^2\mathbb{1}_{\{t\leq\tau\}}\Bigg]ds$$
$$+q(q-1)\frac{\beta_N+\beta_L}{2}\int_0^t \mathbb{E}\Bigg[X_s^{(ij)q-2}\sum_{k,l}$$
$$\left|X_s^{(kj)}\right|\left|\frac{\partial_l\phi_s^k(x)\partial_l\phi_s^i(x)\partial_j\phi_s^i(x)}{\|D\phi_s(x)\|^3}-\frac{\partial_l\phi_s^k(y)\partial_l\phi_s^i(y)\partial_j\phi_s^i(y)}{\|D\phi_s(y)\|^3}\right|\mathbb{1}_{\{t\leq\tau\}}\Bigg]ds$$
$$+\frac{3}{4}q(\beta_N+\beta_L)\int_0^t \mathbb{E}\Bigg[\left|X_s^{(ij)q-1}\right|\sum_{k,l,m,n}\left|\frac{\partial_l\phi_s^k(x)\partial_l\phi_s^n(x)\partial_m\phi_s^k(x)\partial_m\phi_s^n(x)\partial_j\phi_s^i(x)}{\|D\phi_s(x)\|^5}\right.$$
$$-\left.\frac{\partial_l\phi_s^k(y)\partial_l\phi_s^n(y)\partial_m\phi_s^k(y)\partial_m\phi_s^n(y)\partial_j\phi_s^i(y)}{\|D\phi_s(y)\|^5}\right|\mathbb{1}_{\{t\leq\tau\}}\Bigg]ds$$
$$+q(q-1)\int_0^t \mathbb{E}\Bigg[X_s^{(ij)q-2}\sum_{k,l}\left|\frac{\partial_j\phi_s^k(x)\partial_j\phi_s^l(y)}{\|D\phi_s(x)\|\|D\phi_s(y)\|}\right|\left|\partial_k\partial_l b^{i,i}(x_s-y_s)-\partial_k\partial_l b^{i,i}(0)\right|\mathbb{1}_{\{t\leq\tau\}}\Bigg]ds$$
$$+q(q-1)\int_0^t \mathbb{E}\Bigg[X_s^{(ij)q-2}\sum_{k,l,m,n}\left|\frac{\partial_j\phi_s^k(x)\partial_m\phi_s^n(y)\partial_m\phi_s^l(y)\partial_j\phi_s^i(y)}{\|D\phi_s(x)\|\|D\phi_s(y)\|^3}\right.$$
$$+\left.\frac{\partial_j\phi_s^k(y)\partial_m\phi_s^n(x)\partial_m\phi_s^l(x)\partial_j\phi_s^i(x)}{\|D\phi_s(y)\|\|D\phi_s(x)\|^3}\right|\left|\partial_k\partial_n b^{i,l}(x_s-y_s)-\partial_k\partial_n b^{i,l}(0)\right|\mathbb{1}_{\{t\leq\tau\}}\Bigg]ds$$
$$+q(q-1)\int_0^t \mathbb{E}\Bigg[X_s^{(ij)q-2}\sum_{k,l,m,n,p,r}\left|\frac{\partial_l\phi_s^m(x)\partial_l\phi_s^k(x)\partial_j\phi_s^i(x)\partial_p\phi_s^r(y)\partial_p\phi_s^n(y)\partial_j\phi_s^i(y)}{\|D\phi_s(x)\|^3\|D\phi_s(y)\|^3}\right|$$
$$\left|\partial_r\partial_m b^{k,n}(x_s-y_s)-\partial_r\partial_m b^{k,n}(0)\right|\mathbb{1}_{\{t\leq\tau\}}\Bigg]ds$$

7.4 Proof of Lemma 7.2.3: The Two-Point Condition

$$\leq q\mathbb{E}\left[\mathbb{1}_{\{t\leq\tau\}}\check{M}_t^{(ij)}\right]$$

$$+\left|\frac{\beta_L-\beta_N}{4}q^2-\frac{d-1}{2}\beta_N q-\frac{1}{2}\beta_L q\right|\int_0^t \mathbb{E}\left[X_s^{(ij)q}\mathbb{1}_{\{t\leq\tau\}}\right]ds$$

$$+\frac{q(q-1)}{2}\beta_N\int_0^t \mathbb{E}\left[X_s^{(ij)q-2}\sum_k X_s^{(kj)2}\mathbb{1}_{\{t\leq\tau\}}\right]ds$$

$$+q\frac{\beta_N+\beta_L}{2}\int_0^t \mathbb{E}\left[\left|X_s^{(ij)q-1}\right|\sum_{k,l}\left|\frac{\partial_j\phi_s^k(x)\,\partial_l\phi_s^i(x)\,\partial_l\phi_s^k(x)}{\|D\phi_s(x)\|^3}-\frac{\partial_j\phi_s^k(y)\,\partial_l\phi_s^i(y)\,\partial_l\phi_s^k(y)}{\|D\phi_s(y)\|^3}\right|\right.$$

$$\left.\mathbb{1}_{\{t\leq\tau\}}\right]ds+q(q-1)\frac{\beta_N+\beta_L}{4}\int_0^t \mathbb{E}\left[X_s^{(ij)q-2}\sum_{k,n}\right.$$

$$\left.\left(\sum_l \frac{\partial_l\phi_s^n(x)\,\partial_l\phi_s^k(x)\,\partial_j\phi_s^i(x)}{\|D\phi_s(x)\|^3}-\frac{\partial_l\phi_s^n(y)\,\partial_l\phi_s^k(y)\,\partial_j\phi_s^i(y)}{\|D\phi_s(y)\|^3}\right)^2\mathbb{1}_{\{t\leq\tau\}}\right]ds$$

$$+q(q-1)\frac{\beta_N+\beta_L}{2}\int_0^t \mathbb{E}\left[X_s^{(ij)q-2}\sum_{k,l}\right.$$

$$\left.\left|X_s^{(kj)}\right|\left|\frac{\partial_l\phi_s^k(x)\,\partial_l\phi_s^i(x)\,\partial_j\phi_s^i(x)}{\|D\phi_s(x)\|^3}-\frac{\partial_l\phi_s^k(y)\,\partial_l\phi_s^i(y)\,\partial_j\phi_s^i(y)}{\|D\phi_s(y)\|^3}\right|\mathbb{1}_{\{t\leq\tau\}}\right]ds$$

$$+\frac{3}{4}q(\beta_N+\beta_L)\int_0^t \mathbb{E}\left[\left|X_s^{(ij)q-1}\right|\sum_{k,l,m,n}\left|\frac{\partial_l\phi_s^k(x)\,\partial_l\phi_s^n(x)\,\partial_m\phi_s^k(x)\,\partial_m\phi_s^n(x)\,\partial_j\phi_s^i(x)}{\|D\phi_s(x)\|^5}\right.\right.$$

$$\left.\left.-\frac{\partial_l\phi_s^k(y)\,\partial_l\phi_s^n(y)\,\partial_m\phi_s^k(y)\,\partial_m\phi_s^n(y)\,\partial_j\phi_s^i(y)}{\|D\phi_s(y)\|^5}\right|\mathbb{1}_{\{t\leq\tau\}}\right]ds$$

$$+q(q-1)C\int_0^t \mathbb{E}\left[X_s^{(ij)q-2}\sum_{k,l}\left|\frac{\partial_j\phi_s^k(x)\,\partial_j\phi_s^l(y)}{\|D\phi_s(x)\|\,\|D\phi_s(y)\|}\right||x_s-y_s|^2\mathbb{1}_{\{t\leq\tau\}}\right]ds$$

$$+q(q-1)C\int_0^t \mathbb{E}\left[X_s^{(ij)q-2}\sum_{k,l,m,n}\left|\frac{\partial_j\phi_s^k(x)\,\partial_m\phi_s^n(y)\,\partial_m\phi_s^l(y)\,\partial_j\phi_s^i(y)}{\|D\phi_s(x)\|\,\|D\phi_s(y)\|^3}\right.\right.$$

$$\left.\left.+\frac{\partial_j\phi_s^k(y)\,\partial_m\phi_s^n(x)\,\partial_m\phi_s^l(x)\,\partial_j\phi_s^i(x)}{\|D\phi_s(y)\|\,\|D\phi_s(x)\|^3}\right||x_s-y_s|^2\mathbb{1}_{\{t\leq\tau\}}\right]ds$$

$$+q(q-1)C\int_0^t \mathbb{E}\left[X_s^{(ij)q-2}\sum_{k,l,m,n,p,r}\left|\frac{\partial_l\phi_s^m(x)\,\partial_l\phi_s^k(x)\,\partial_j\phi_s^i(x)\,\partial_p\phi_s^r(y)\,\partial_p\phi_s^n(y)\,\partial_j\phi_s^i(y)}{\|D\phi_s(x)\|^3\,\|D\phi_s(y)\|^3}\right|\right.$$

$$\left.|x_s-y_s|^2\mathbb{1}_{\{t\leq\tau\}}\right]ds.$$

Recall the inequality $\prod a_i - \prod b_i \leq \sum_i |a_i - b_i| \leq d\max_i |a_i - b_i|$ (that is valid for numbers a_i, b_i with $|a_i| \vee |b_i| \leq 1$) and use that the expectation of a positive random variable on an event is growing in the event and conclude that the latter is less or equal to

$$q\mathbb{E}\left[\mathbb{1}_{\{t\leq\tau\}}\check{M}_t^{(ij)}\right]$$
$$+\left|\frac{\beta_L-\beta_N}{4}q^2-\frac{d-1}{2}\beta_N q-\frac{1}{2}\beta_L q\right|\int_0^t \mathbb{E}\left[\max_{i,j} X_s^{(ij)q}\mathbb{1}_{\{t\leq\tau\}}\vee |x_s-y_s|^q\mathbb{1}_{\{t\leq\tau\}}\right]ds$$
$$+\frac{q(q-1)}{2}\beta_N d\int_0^t \mathbb{E}\left[\max_{i,j} X_s^{(ij)q}\mathbb{1}_{\{t\leq\tau\}}\vee |x_s-y_s|^q\mathbb{1}_{\{t\leq\tau\}}\right]ds$$
$$+q\frac{\beta_N+\beta_L}{2}3d^2\int_0^t \mathbb{E}\left[\max_{i,j} X_s^{(ij)q}\mathbb{1}_{\{t\leq\tau\}}\vee |x_s-y_s|^q\mathbb{1}_{\{t\leq\tau\}}\right]ds$$
$$+q(q-1)\frac{\beta_N+\beta_L}{4}9d^4\int_0^t \mathbb{E}\left[\max_{i,j} X_s^{(ij)q}\mathbb{1}_{\{t\leq\tau\}}\vee |x_s-y_s|^q\mathbb{1}_{\{t\leq\tau\}}\right]ds$$
$$+q(q-1)\frac{\beta_N+\beta_L}{2}3d^2\int_0^t \mathbb{E}\left[\max_{i,j} X_s^{(ij)q}\mathbb{1}_{\{t\leq\tau\}}\vee |x_s-y_s|^q\mathbb{1}_{\{t\leq\tau\}}\right]ds$$
$$+\frac{3}{4}q(\beta_N+\beta_L)5d^4\int_0^t \mathbb{E}\left[\max_{i,j} X_s^{(ij)q}\mathbb{1}_{\{t\leq\tau\}}\vee |x_s-y_s|^q\mathbb{1}_{\{t\leq\tau\}}\right]ds$$
$$+q(q-1)(d^2+2d^4+d^6)C\int_0^t \mathbb{E}\left[\max_{i,j} X_s^{(ij)q}\mathbb{1}_{\{t\leq\tau\}}\vee |x_s-y_s|^q\mathbb{1}_{\{t\leq\tau\}}\right]ds$$
$$\leq q\mathbb{E}\left[\mathbb{1}_{\{t\leq\tau\}}\check{M}_t^{(ij)}\right]$$
$$+\left|\frac{\beta_L-\beta_N}{4}q^2-\frac{d-1}{2}\beta_N q-\frac{1}{2}\beta_L q\right|\int_0^t \mathbb{E}\left[\left(\max_{i,j} X_s^{(ij)q}\vee |x_s-y_s|^q\right)\mathbb{1}_{\{s\leq\tau\}}\right]ds$$
$$+\frac{q(q-1)}{2}\beta_N d\int_0^t \mathbb{E}\left[\left(\max_{i,j} X_s^{(ij)q}\vee |x_s-y_s|^q\right)\mathbb{1}_{\{s\leq\tau\}}\right]ds$$
$$+q\frac{\beta_N+\beta_L}{2}3d^2\int_0^t \mathbb{E}\left[\left(\max_{i,j} X_s^{(ij)q}\vee |x_s-y_s|^q\right)\mathbb{1}_{\{s\leq\tau\}}\right]ds$$
$$+q(q-1)\frac{\beta_N+\beta_L}{4}9d^4\int_0^t \mathbb{E}\left[\left(\max_{i,j} X_s^{(ij)q}\vee |x_s-y_s|^q\right)\mathbb{1}_{\{s\leq\tau\}}\right]ds$$
$$+q(q-1)\frac{\beta_N+\beta_L}{2}3d^2\int_0^t \mathbb{E}\left[\left(\max_{i,j} X_s^{(ij)q}\vee |x_s-y_s|^q\right)\mathbb{1}_{\{s\leq\tau\}}\right]ds$$
$$+\frac{3}{4}q(\beta_N+\beta_L)5d^4\int_0^t \mathbb{E}\left[\left(\max_{i,j} X_s^{(ij)q}\vee |x_s-y_s|^q\right)\mathbb{1}_{\{s\leq\tau\}}\right]ds$$
$$+q(q-1)(d^2+2d^4+d^6)C\int_0^t \mathbb{E}\left[\left(\max_{i,j} X_s^{(ij)q}\vee |x_s-y_s|^q\right)\mathbb{1}_{\{s\leq\tau\}}\right]ds$$
$$=q\mathbb{E}\left[\mathbb{1}_{\{t\leq\tau\}}\check{M}_t^{(ij)}\right]+\kappa_q\int_0^t \mathbb{E}\left[\left(\max_{i,j} X_s^{(ij)q}\vee |x_s-y_s|^q\right)\mathbb{1}_{\{s\leq\tau\}}\right]ds.$$

This enables us to conclude with Proposition 7.4.7 and Lemmas 1.3.8 and 1.4.3 in the

7.4 Proof Of Lemma 7.2.3: The Two-Point Condition

following way.

$$\mathbb{E}\left[\max_{i,j} X_t^{(ij)q}\mathbb{1}_{\{t\leq\tau\}} \vee |x_t - y_t|^q \mathbb{1}_{\{t\leq\tau\}}\right]$$

$$\leq \sum_{i,j} \mathbb{E}\left[X_t^{(ij)q}\mathbb{1}_{\{t\leq\tau\}}\mathbb{1}_{\{t\leq\tau\}}\right] + \mathbb{E}\left[|x_t-y_t|^q\right] \leq \mathbb{E}\left[|x_t-y_t|^q\right]$$

$$+ d^2 q \sum_{i,j}\mathbb{E}\left[\mathbb{1}_{\{t\leq\tau\}}\check{M}_t^{(ij)}\right] + d^2\kappa_q \int_0^t \mathbb{E}\left[\left(\max_{i,j} X_s^{(ij)q} \vee |x_s-y_s|^q\right)\mathbb{1}_{\{s\leq\tau\}}\right]ds$$

$$\leq d^2 q |x-y|^{2q} \frac{\bar{c}_3}{\tilde{r}^{2q}} e^{(\Lambda_3+\frac{1}{2}\sigma_3^2 q+\sigma_4^2 q^2)t} + 2^q |x-y|^q e^{(\lambda q+\frac{1}{2}q^2\bar{\sigma}^2)t}$$

$$+ d^2 \kappa_q \int_0^t \mathbb{E}\left[\left(\max_{i,j} X_s^{(ij)q} \vee |x_s-y_s|^q\right)\mathbb{1}_{\{s\leq\tau\}}\right]ds$$

$$\leq d^2 \kappa_q \int_0^t \mathbb{E}\left[\left(\max_{i,j} X_s^{(ij)q} \vee |x_s-y_s|^q\right)\mathbb{1}_{\{s\leq\tau\}}\right]ds$$

$$+ |x-y|^q \left(\frac{2\bar{c}_3 d^2 q}{\tilde{r}^{2q}}\right) e^{(\Lambda_3+(\lambda\vee\frac{1}{2}\sigma_3^2)q+(\frac{1}{2}\bar{\sigma}^2\vee\sigma_4^2)q^2)t}$$

since we assumed $\tilde{r} \leq 2^{-1/2}$. This now implies via Grönwall's inequality (see [51, I.§2.1] for an appropriate version)

$$h(t) \leq |x-y|^q 2\frac{\bar{c}_3 d^2 q}{\tilde{r}^{2q}} \frac{e^{(\Lambda_3+(\lambda\vee\frac{1}{2}\sigma_3^2)q+(\frac{1}{2}\bar{\sigma}^2\vee\sigma_4^2)q^2+d^2\kappa_q)t}}{\Lambda_3+(\lambda\vee\frac{1}{2}\sigma_3^2)q+(\frac{1}{2}\bar{\sigma}^2\vee\sigma_4^2)q^2+d^2\kappa_q}.$$

The proof is complete. □

The next lemma will be the last ingredient to the proof of Lemma 7.2.3.

Lemma 7.4.9 (second estimate for τ after T).
We have for even $q \geq \frac{4\sqrt{\beta_L}[2(d-1)\beta_N-c-2\beta_L]}{128\beta_L} \vee 3$ that

$$\mathbb{E}\left[\sup_{0\leq t\leq T}|\psi_t(x)-\psi_t(y)|^q\mathbb{1}_{\{T\leq\tau\}}\right]^{\frac{1}{q}} \leq \bar{c}_6\sqrt{q}|x-y|e^{(\Lambda_6+\frac{1}{2}\sigma_2^2 q)T}$$

with $\bar{c}_6 = \sup_{q\geq 3}\left(\frac{\bar{c}_5^{\frac{1}{q}}\sqrt{2}d^2 C_q^{\frac{1}{q}}\sqrt{\beta_L+\beta_N}+\sqrt{C}d^3 C_q^{\frac{1}{q}}\sqrt{2}+2d^4(\beta_N+\beta_L)}{(\Lambda_5 q+\sigma_5^2 q^2)^{\frac{1}{q}}\sqrt{q}}\right)$, $\Lambda_6 = \Lambda_5 + \frac{1}{e}$ and $\sigma_6 := \sqrt{2}\sigma_5$.

Remark: Observe that by Lemma 7.1.4 c_6 is indeed finite.

Proof: By the triangle inequality and Lemmas 7.1.4 and 7.4.1 we have

$$\mathbb{E}\left[\sup_{0\leq t\leq T}|\psi_t(x)-\psi_t(y)|^q \mathbb{1}_{\{T\leq\tau\}}\right]^{\frac{1}{q}} \leq \mathbb{E}\left[\sup_{0\leq t\leq T}\tilde{M}_t^q \mathbb{1}_{\{T\leq\tau\}}\right]^{\frac{1}{q}} + \mathbb{E}\left[\sup_{0\leq t\leq T}A_t^q \mathbb{1}_{\{T\leq\tau\}}\right]^{\frac{1}{q}}$$

$$\leq \mathbb{E}\left[\sup_{0\leq t\leq T}\tilde{M}_{t\wedge\tau}^q \mathbb{1}_{\{T\leq\tau\}}\right]^{\frac{1}{q}} + \mathbb{E}\left[\sup_{0\leq t\leq T}A_{t\wedge\tau}^q\right]^{\frac{1}{q}} \leq C_q^{\frac{1}{q}}\mathbb{E}\left[\langle M\rangle_{T\wedge\tau}^{\frac{q}{2}}\right]^{\frac{1}{q}} + \mathbb{E}\left[A_{T\wedge\tau}^q\right]^{\frac{1}{q}}$$

$$\leq C_q^{\frac{1}{q}}\mathbb{E}\left[\left(\frac{\beta_L+\beta_N}{2}\int_0^{T\wedge\tau}\sum_{i,k}\left(\sum_j \frac{\partial_j\phi_s^i(x)\partial_j\phi_s^k(x)}{\|D\phi_s(x)\|^2}-\frac{\partial_j\phi_s^i(y)\partial_j\phi_s^k(y)}{\|D\phi_s(y)\|^2}\right)^2 ds\right.\right.$$

$$+2\sum_{i,j,k,l,m,n}\int_0^{T\wedge\tau}\frac{\partial_j\phi_s^i(x)\partial_j\phi_s^k(x)\partial_m\phi_s^l(y)\partial_m\phi_s^n(y)}{\|D\phi_s(x)\|^2\|D\phi_s(y)\|^2}\left[\partial_k\partial_n b^{i,l}(x_s-y_s)-\partial_k\partial_n b^{i,l}(0)\right]ds\bigg)^{\frac{q}{2}}\bigg]^{\frac{1}{q}}$$

$$+\mathbb{E}\left[\left(\frac{(\beta_L+\beta_N)}{2}\int_0^{T\wedge\tau}\sum_{i,j,k,m}\frac{\partial_j\phi_s^k(y)\partial_j\phi_s^i(y)\partial_m\phi_s^k(y)\partial_m\phi_s^i(y)}{\|D\phi_s(y)\|^4}\right.\right.$$

$$\left.\left.-\frac{\partial_j\phi_s^k(x)\partial_j\phi_s^i(x)\partial_m\phi_s^k(x)\partial_m\phi_s^i(x)}{\|D\phi_s(x)\|^4}ds\right)^q\right]^{\frac{1}{q}}$$

$$\leq C_q^{\frac{1}{q}}\sqrt{\frac{\beta_L+\beta_N}{2}}\mathbb{E}\left[\left(\int_0^{T\wedge\tau}\sum_{i,k}\left(\sum_j \frac{\partial_j\phi_s^i(x)\partial_j\phi_s^k(x)}{\|D\phi_s(x)\|^2}-\frac{\partial_j\phi_s^i(y)\partial_j\phi_s^k(y)}{\|D\phi_s(y)\|^2}\right)^2 ds\right)^{\frac{q}{2}}\right]^{\frac{1}{q}}$$

$$+C_q^{\frac{1}{q}}\sqrt{2}\mathbb{E}\left[\sum_{i,j,k,l,m,n}\int_0^{T\wedge\tau}\frac{\partial_j\phi_s^i(x)\partial_j\phi_s^k(x)\partial_m\phi_s^l(y)\partial_m\phi_s^n(y)}{\|D\phi_s(x)\|^2\|D\phi_s(y)\|^2}\right.$$

$$\left.\left[\partial_k\partial_n b^{i,l}(x_s-y_s)-\partial_k\partial_n b^{i,l}(0)\right]ds^{\frac{q}{2}}\right]^{\frac{1}{q}}$$

$$+\frac{\beta_N+\beta_L}{2}\mathbb{E}\left[\left(\int_0^{T\wedge\tau}\sum_{i,j,k,m}\frac{\partial_j\phi_s^k(y)\partial_j\phi_s^i(y)\partial_m\phi_s^k(y)\partial_m\phi_s^i(y)}{\|D\phi_s(y)\|^4}\right.\right.$$

$$\left.\left.-\frac{\partial_j\phi_s^k(x)\partial_j\phi_s^i(x)\partial_m\phi_s^k(x)\partial_m\phi_s^i(x)}{\|D\phi_s(x)\|^4}ds\right)^q\right]^{\frac{1}{q}}$$

$$\leq C_q^{\frac{1}{q}}\sqrt{\frac{\beta_L+\beta_N}{2}}\mathbb{E}\left[\left(\int_0^{T\wedge\tau} 4d^4\left(\max_{i,j}X_t^{(ij)}\vee|x_t-y_t|\right)^2 ds\right)^{\frac{q}{2}}\right]^{\frac{1}{q}}$$

$$+C_q^{\frac{1}{q}}\sqrt{2}\mathbb{E}\left[\left(Cd^6\int_0^{T\wedge\tau}\left(\max_{i,j}X_t^{(ij)}\vee|x_t-y_t|\right)^2 ds\right)^{\frac{q}{2}}\right]^{\frac{1}{q}}$$

$$+\frac{\beta_N+\beta_L}{2}\mathbb{E}\left[\left(4d^4\int_0^{T\wedge\tau}\max_{i,j}X_t^{(ij)}\vee|x_t-y_t|ds\right)^q\right]^{\frac{1}{q}}.$$

7.4 Proof Of Lemma 7.2.3: The Two-Point Condition

By Jensen's inequality this is less or equal to

$$2d^2 C_q^{\frac{1}{q}} \sqrt{\frac{\beta_L + \beta_N}{2}} \mathbb{E}\left[(T \wedge \tau)^{\frac{q}{2}-1} \int_0^{T\wedge\tau} \left(\max_{i,j} X_t^{(ij)} \vee |x_t - y_t|\right)^q \mathbb{1}_{\{T \leq \tau\}} ds\right]^{\frac{1}{q}}$$

$$+ \sqrt{C} d^3 C_q^{\frac{1}{q}} \sqrt{2} \mathbb{E}\left[(T \wedge \tau)^{\frac{q}{2}-1} \int_0^{T\wedge\tau} \left(\max_{i,j} X_t^{(ij)} \vee |x_t - y_t|\right)^q \mathbb{1}_{\{T \leq \tau\}} ds\right]^{\frac{1}{q}}$$

$$+ 4d^4 \frac{\beta_N + \beta_L}{2} \mathbb{E}\left[(T \wedge \tau)^{q-1} \int_0^{T\wedge\tau} \left(\max_{i,j} X_t^{(ij)} \vee |x_t - y_t|\right)^q \mathbb{1}_{\{T \leq \tau\}} ds\right]^{\frac{1}{q}}$$

$$\leq \sqrt{2} d^2 C_q^{\frac{1}{q}} \sqrt{\beta_L + \beta_N} T^{\frac{q-2}{2q}} \mathbb{E}\left[\int_0^T \left(\max_{i,j} X_t^{(ij)} \vee |x_t - y_t|\right)^q \mathbb{1}_{\{t \leq \tau\}} ds\right]^{\frac{1}{q}}$$

$$+ \sqrt{C} d^3 C_q^{\frac{1}{q}} \sqrt{2} T^{\frac{q-2}{2q}} \mathbb{E}\left[\int_0^T \left(\max_{i,j} X_t^{(ij)} \vee |x_t - y_t|\right)^q \mathbb{1}_{\{t \leq \tau\}} ds\right]^{\frac{1}{q}}$$

$$+ 2d^4 (\beta_N + \beta_L) T^{\frac{q-1}{q}} \mathbb{E}\left[\int_0^T \left(\max_{i,j} X_t^{(ij)} \vee |x_t - y_t|\right)^q \mathbb{1}_{\{t \leq \tau\}} ds\right]^{\frac{1}{q}}.$$

Another application of Fubini's theorem and the Lemmas 7.1.1 and 7.4.8 now yields that the latter is less or equal to (remember that we chose even $q \geq 3$)

$$\left(\sqrt{2} d^2 C_q^{\frac{1}{q}} \sqrt{\beta_L + \beta_N} T^{\frac{q-2}{2q}} + \sqrt{C} d^3 C_q^{\frac{1}{q}} \sqrt{2} T^{\frac{q-2}{2q}} + 2d^4 (\beta_N + \beta_L) T^{\frac{q-1}{q}}\right)\left(\int_0^T h(s) ds\right)^{\frac{1}{q}}$$

$$\leq \left(\sqrt{2} d^2 C_q^{\frac{1}{q}} \sqrt{\beta_L + \beta_N} + \sqrt{C} d^3 C_q^{\frac{1}{q}} \sqrt{2} + 2d^4 (\beta_N + \beta_L)\right) e^{\frac{T}{e}} \left(\int_0^T \bar{c}_5 \tilde{r}^{-2q} |x-y|^q e^{(\Lambda_5 q + \sigma_5^2 q^2)t}\right)^{\frac{1}{q}}$$

$$= \frac{\bar{c}_5^{\frac{1}{q}}}{\tilde{r}^2} |x-y| \left(\sqrt{2} d^2 C_q^{\frac{1}{q}} \sqrt{\beta_L + \beta_N} + \sqrt{C} d^3 C_q^{\frac{1}{q}} \sqrt{2} + 2d^4 (\beta_N + \beta_L)\right) e^{\frac{T}{e}} \left(\frac{e^{(\Lambda_5 q + \sigma_5^2 q^2)T} - 1}{\Lambda_5 q + \sigma_5^2 q^2}\right)^{\frac{1}{q}}$$

$$\leq \frac{\bar{c}_5^{\frac{1}{q}}}{\tilde{r}^2} |x-y| \left(\sqrt{2} d^2 C_q^{\frac{1}{q}} \sqrt{\beta_L + \beta_N} + \sqrt{C} d^3 C_q^{\frac{1}{q}} \sqrt{2} + 2d^4 (\beta_N + \beta_L)\right) e^{\frac{T}{e}} \frac{e^{(\Lambda_5 + \sigma_5^2 q)T}}{(\Lambda_5 q + \sigma_5^2 q^2)^{\frac{1}{q}}}$$

$$\leq \left(\sqrt{2} d^2 C_q^{\frac{1}{q}} \sqrt{\beta_L + \beta_N} + \sqrt{C} d^3 C_q^{\frac{1}{q}} \sqrt{2} + 2d^4 (\beta_N + \beta_L)\right) \frac{\bar{c}_5^{\frac{1}{q}} |x-y| e^{(\Lambda_5 + \frac{1}{e} + \sigma_5^2 q)T}}{(\Lambda_5 q + \sigma_5^2 q^2)^{\frac{1}{q}} \tilde{r}^2}.$$

The proof is complete. □

We fix now $\tilde{r} \leq \bar{r} \wedge 2^{-1/2}$ subject to (7.14), $r \leq e^{-1} \tilde{r}$ and conclude that since we have that $\sup_{0 \leq t \leq T} |\psi_t(x) - \psi_t(y)|^q$ is the sum of the two terms $\sup_{0 \leq t \leq T} |\psi_t(x) - \psi_t(y)|^q \mathbb{1}_{\{T \leq \tau\}}$ and $\sup_{0 \leq t \leq T} |\psi_t(x) - \psi_t(y)|^q \mathbb{1}_{\{T > \tau\}}$ another application of the triangle inequality together with

Lemmas 7.4.2, 7.4.6 and 7.4.9 completes the proof of Lemma 7.2.3. □

7.5 A New Proof For Linear Expansion

We include a short proof for the fact that exponential growth of the first order derivative of a stochastic flows implies linear growth of the diameter (with suprema taken over compact sets in both cases as before). The idea of the proof is originally due to T. Lyons and was communicated by M. Scheutzow. Nevertheless all remaining mistakes are mine.

Lemma 7.5.1 (exponential growth and linear growth)**.**
Let ϕ be a stochastic flow such that the one-point motions $x_t = \phi_t(x)$ satisfy for $R > 0$ that

$$\max_{|x| \leq R} \mathbb{P}\left[\max_{0 \leq t \leq T} |x_t| \geq CT - 1\right] \leq e^{-f(C)T} \text{ with } \lim_{C \to \infty} f(C) = \infty.$$

Assume further that there exists a constant $K \geq 0$ such that for any $R > 0$ we have

$$\limsup_{t \to \infty} \frac{1}{t} \sup_{|x| \leq R} \log \|D\phi_t(x)\| \leq K \text{ a.s..}$$

Then there exists $C > 0$ such that we have for any $R > 0$ that

$$\limsup_{t \to \infty} \frac{1}{t} \sup_{|x| \leq R} |\phi_t(x)| \leq C \text{ a.s..}$$

Proof: The covering of $K_R(x)$ with balls of radius e^{-rT} centered in $x^{(1)}, \ldots, x^{(N)}$ can be done with at most $N := ce^{rdT}$ balls for some constant c. Hence we may conclude with the mean value theorem

$$\{\frac{1}{T} \sup_{|x| \leq R} |\phi_T(x)| \geq C\} \subset \{|x_T^{(i)}| \geq CT - 1 \text{ for some } i \in \{1, \ldots, N\}\}$$

$$\cup \{\text{diam}(K_{e^{-rT}}(x_T^{(i)})) \geq 1 \text{ for some } i \in \{1, \ldots, N\}\}$$

$$\subset \{|x_T^{(i)}| \geq CT - 1 \text{ for some } i \in \{1, \ldots, N\}\} \cup \{\sup_{|x| \leq R} \log \|D\phi_T(x)\| > rT\}$$

$$:= F_1(T) \cup F_2(T).$$

7.5 A New Proof For Linear Expansion

Since

$$\mathbb{P}\left[F_1(T)\right] = \mathbb{P}\left[|x_T^{(i)}| \geq CT - 1 \text{ for some } i \in \{1, \ldots, N\}\right]$$
$$\leq ce^{rdT} \max_{|x| \leq R} \mathbb{P}\left[|x_T| \geq CT - 1\right] \leq ce^{rdT} e^{-f(C)T}$$

we get for sufficiently large C that $F_1(T)$ occurs only at finitely many $T \in \mathbb{N}$ by the first lemma of Borel and Cantelli. If we increase C a bit more then this holds also for general $T \in \mathbb{R}$. Hence there is a.s. a $T > 0$ such that for $t \geq T$ have that $\Omega \setminus \cap_{t \geq T} F_1(t)$ occurs. Since by assumption the same is true for $F_2(t)$ we get that if we choose $r > K$ the proof is complete. □

Remark: IBFs and IOUFs satisfy the condition on the one-point motion given in the lemma. RIFs do not.

Chapter 8

Asymptotic Growth Of Spatial Derivatives: General Case

We generalize the results from Chapter 7 to a much more general framework. Again the central tool for the proof is Theorem 7.1.3. Nevertheless we need different methods to verify its assumptions than those used in Chapter 7. This chapter is joint work with M. Scheutzow.

8.1 The Main Result

The main result looks roughly like the one of Chapter 7 except that we have to specify the conditions on the local characteristics a and b and the fact that we use $\log(1+\cdot)$ instead of $\log(\cdot)$. Note that we leave the scope of isotropic flows so b will denote the drift part of the general flow SDE and not an isotropic covariance tensor. Let us recall the setup from Chapter 1. Let $(\phi_{s,t}: 0 \leq s \leq t < \infty)$ be a stochastic flow on $\mathbb{R}^d$ obtained as the solution of (1.2). We recall that for some m we have the decomposition $F(t,x) = M(t,x) + V(t,x)$ where $M(t,x)$ is a $C^{m,\delta}$-local martingale and $V(t,x)$ is a continuous $C^{m,\delta}$-process, such that $D_x^\alpha V(t,x)$, $|\alpha| \leq m$ are all processes of bounded variation. Assume again that the local characteristic (a,b) can be written as

$$\langle M(\cdot,x), M(\cdot,y)\rangle_t = \int_0^t a(x,y,s)ds \quad \text{and} \quad V(t,x) = \int_0^t b(x,s)ds$$

wherein $a(x,y,s) = (a^{ij}(x,y,s))_{1\leq i,j \leq d}$ and $b(x,s) = b^i(x,s)_{1\leq i \leq d}$ are as in Chapter 1. We will use α, β, γ and δ to denote multi-indices in this chapter. With these notations we can state the following theorem. Note that IOUFs, IBFs and RIFs are covered by it.

Theorem 8.1.1 (exponential growth of spatial derivatives).
Assume $(a,b) \in B_{ub}^{n,1}$, let Ξ be a compact subset of $\mathbb{R}^d$ with box dimension $\Delta > 0$ and let α be a multi-index with $|\alpha| \leq n$. Then

$$\limsup_{T \to \infty} \left(\sup_{0 \leq t \leq T} \sup_{x \in \Xi} \frac{1}{T} \log(1 + |D^\alpha \phi_t(x)|) \right) \leq K \text{ a.s.}$$

where we put

$$K := \begin{cases} B_n + 2A_n \sigma_n \Delta & : \text{if } 0 \geq \Lambda_0 \\ B_n + A_n d \sigma_n \sqrt{\frac{\Delta}{d - \Delta}} & : \text{otherwise} \end{cases} \quad \text{and} \quad \Lambda_0 := \frac{\sigma_n^2 d}{\Delta}(\frac{d}{2} - \Delta)$$

for constants A_n, B_n and σ_n that depend only on n, d, (a,b).

The proof of this theorem relies on two lemmas as before. These allow us to apply Theorem 7.1.3 to $\psi_t(x) := \log(1 + |D^\alpha \phi_t^j(x)|)$. $\square$

Lemma 8.1.2 (condition on the one-point motion).
Assume that $(a,b) \in B_{ub}^{n,1}$. We have for each bounded $S \subset \mathbb{R}^d$ that

$$\limsup_{T \to \infty} \frac{1}{T} \log \sup_{x \in S} \mathbb{P}\left[\sup_{0 \leq t \leq T} \sum_{1 \leq |\alpha| \leq n, j} \log(1 + |D^\alpha \phi_t^j(x)|) > kT \right] \leq -\frac{(k - B_n)_+^2}{2A_n^2}.$$

Therein A_n and B_n are constants that depend only on n, d and (a,b) as in Theorem 8.1.1.

Remark: In fact we will show that we can take the supremum over $\mathbb{R}^d$ instead of over bounded S.

Lemma 8.1.3 (condition on the two-point motion).
Assume that $(a,b) \in B_{ub}^{n,1}$. We have for each $x, y \in \mathbb{R}^d$, $T > 0$ and $p \geq 1$ that
$\mathbb{E}\left[\sup_{0 \leq t \leq T} |\log(1 + |D^\alpha \phi_x^j(t)|) - \log(1 + |D^\alpha \phi_y^j(t)|)|^p \right]^{1/p} \leq \bar{c} |x - y| e^{\frac{1}{2} p \sigma_n^2 T}$ *for σ_n as in Theorem 7.2.1 (i.e. we specify them later).*

8.2 Proof Of Lemma 8.1.2: The One-Point Condition

First recall from [20, (18)] that we have for $x \in \mathbb{R}^d$, arbitrary α with $1 \leq |\alpha| \leq n$ and $1 \leq j \leq d$ that

$$D^\alpha \phi_t^j(x) = \eta_{\alpha,j} + \sum_i \int_0^t D^\alpha \phi_s^i(x) \, \partial_i F^j(ds, x_s) + \sum_{2 \leq |\beta| \leq |\alpha|} \int_0^t P_{\beta,x,j}(s) D^\beta F^j(ds, x_s) \quad (8.1)$$

where $\eta_{\alpha,j} = 1$ if $|\alpha| = 1$ and $\alpha_j = 1$ and $\eta_{\alpha,j} = 0$ otherwise, and where $P_{\beta,x,j}(s)$ is a finite sum of products of the form $\prod_{i=1}^r D^{\gamma_i} \phi_{0s}^{j_i}(x)$ with $1 \leq |\gamma_i| \leq |\alpha|$ and $\sum_{i=1}^r |\gamma_i| = \alpha$. Observe that the last sum in (8.1) is empty if $|\alpha| = 1$. (8.1) yields with Itô's formula

$$(D^\alpha \phi_t^j(x))^2$$
$$= \eta_{\alpha,j} + 2\sum_i \int_0^t D^\alpha \phi_s^j(x) D^\alpha \phi_s^i(x) \, \partial_i M^j(ds, x_s)$$
$$+ 2 \sum_{2 \leq |\beta| \leq |\alpha|} \int_0^t D^\alpha \phi_s^j(x) P_{\beta,x,j}(s) D^\beta M^j(ds, x_s)$$
$$+ \sum_{i,k} \int_0^t D^\alpha \phi_s^i(x) D^\alpha \phi_s^k(x) D_1^i D_2^k a^{jj}(x_s, x_s, s) ds$$
$$+ \sum_{2 \leq |\beta|, |\gamma| \leq |\alpha|} P_{\beta,x,j}(s) P_{\gamma,x,j}(s) D_1^\beta D_2^\gamma a^{jj}(x_s, x_s, s) ds$$
$$+ 2 \sum_{2 \leq |\beta| \leq |\alpha|, i} \int_0^t D^\alpha \phi_s^i(x) P_{\beta,x,j}(s) D_1^i D_2^\beta a^{jj}(x_s, x_s, s) ds$$
$$+ 2\sum_i \int_0^t D^\alpha \phi_s^j(x) D^\alpha \phi_s^i(x) D^i b^j(x_s, s) ds + 2 \sum_{2 \leq |\beta| \leq |\alpha|} \int_0^t D^\alpha \phi_s^j(x) P_{\beta,x,j}(s) D^\beta b^i(x_s, s) ds$$

and hence summing over α and j that

$$\psi_t^{(n)}(x) := \sum_{1 \leq |\alpha| \leq n, j} (D^\alpha \phi_t^j(x))^2$$
$$= d + 2 \sum_{1 \leq |\alpha| \leq n, i, j} \int_0^t D^\alpha \phi_s^j(x) D^\alpha \phi_s^i(x) \, \partial_i M^j(ds, x_s)$$
$$+ 2 \sum_{2 \leq |\beta| \leq |\alpha| \leq n, j} \int_0^t D^\alpha \phi_s^j(x) P_{\beta,x,j}(s) D^\beta M^j(ds, x_s)$$

$$+ \sum_{1\leq|\alpha|\leq n, i,j,k} \int_0^t D^\alpha \phi_s^i(x)\, D^\alpha \phi_s^k(x)\, D_1^i D_2^k a^{jj}(x_s, x_s, s)\, ds$$

$$+ \sum_{2\leq|\beta|,|\gamma|\leq|\alpha|\leq n, j} \int_0^t P_{\beta,x,j}(s)\, P_{\gamma,x,j}(s)\, D_1^\beta D_2^\gamma a^{jj}(x_s, x_s, s)\, ds$$

$$+ 2 \sum_{2\leq|\beta|\leq|\alpha|\leq n, i, j} \int_0^t D^\alpha \phi_s^i(x)\, P_{\beta,x,j}(s)\, D_1^i D_2^\beta a^{jj}(x_s, x_s, s)\, ds$$

$$+ 2 \sum_{1\leq|\alpha|\leq n, i, j} \int_0^t D^\alpha \phi_s^j(x)\, D^\alpha \phi_s^i(x)\, D^i b^j(x_s, s)\, ds$$

$$+ 2 \sum_{2\leq|\beta|\leq|\alpha|\leq n, j} \int_0^t D^\alpha \phi_s^j(x)\, P_{\beta,x,j}(s)\, D^\beta b^i(x_s, s)\, ds.$$

Note that $\sum_{2\leq|\beta|\leq|\alpha|\leq n, j}$ means that the sum is taken over $j \in \{1,\ldots,d\}$ and all multi-indices α and β with $1 \leq |\beta| \leq |\alpha| \leq n$. Applying the shifted logarithm $\log(1 + \cdot)$ we get with another application of Itô's formula that

$$\log\left(1 + \psi_t^{(n)}(x)\right) = \log(1 + d) + 2 \sum_{1\leq|\alpha|\leq n, i, j} \int_0^t \frac{D^\alpha \phi_s^j(x)\, D^\alpha \phi_s^i(x)}{1 + \psi_s^{(n)}} \partial_i M^j(ds, x_s)$$

$$+ 2 \sum_{2\leq|\beta|\leq|\alpha|\leq n, j} \int_0^t \frac{D^\alpha \phi_s^j(x)\, P_{\beta,x,j}(s)}{1 + \psi_s^{(n)}} D^\beta M^j(ds, x_s)$$

$$+ \sum_{1\leq|\alpha|\leq n, i, j, k} \int_0^t \frac{D^\alpha \phi_s^i(x)\, D^\alpha \phi_s^k(x)}{1 + \psi_s^{(n)}} D_1^i D_2^k a^{jj}(x_s, x_s, s)\, ds$$

$$+ \sum_{2\leq|\beta|,|\gamma|\leq|\alpha|\leq n, j} \int_0^t \frac{P_{\beta,x,j}(s)\, P_{\gamma,x,j}(s)}{1 + \psi_s^{(n)}} D_1^\beta D_2^\gamma a^{jj}(x_s, x_s, s)\, ds$$

$$+ 2 \sum_{2\leq|\beta|\leq|\alpha|\leq n, i, j} \int_0^t \frac{D^\alpha \phi_s^i(x)\, P_{\beta,x,j}(s)}{1 + \psi_s^{(n)}} D_1^i D_2^\beta a^{jj}(x_s, x_s, s)\, ds$$

$$+ 2 \sum_{1\leq|\alpha|\leq n, i, j} \int_0^t \frac{D^\alpha \phi_s^j(x)\, D^\alpha \phi_s^i(x)}{1 + \psi_s^{(n)}} D^i b^j(x_s, s)\, ds$$

$$+ 2 \sum_{2\leq|\beta|\leq|\alpha|\leq n, j} \int_0^t \frac{D^\alpha \phi_s^j(x)\, P_{\beta,x,j}(s)}{1 + \psi_s^{(n)}} D^\beta b^i(x_s, s)\, ds$$

$$- 4 \sum_{1\leq|\alpha|,|\delta|\leq n, i, j, k, l} \frac{D^\alpha \phi_s^j(x)\, D^\alpha \phi_s^i(x)\, D^\delta \phi_s^l(x)\, D^\delta \phi_s^k(x)}{(1 + \psi_s^{(n)})^2} D_1^i D_2^k a^{jl}(x_s, x_s, s)\, ds$$

$$- 4 \sum_{2\leq|\beta|\leq|\alpha|\leq n,\, 2\leq|\gamma|\leq|\delta|\leq n, i, j} \int_0^t \frac{D^\alpha \phi_s^j(x)\, P_{\beta,x,j}(s)\, D^\delta \phi_s^i(x)\, P_{\gamma,x,i}(s)}{(1 + \psi_s^{(n)})^2} D_1^\beta D_2^\gamma a^{ji}(x_s, x_s, s)\, ds$$

8.2 Proof Of Lemma 8.1.2: The One-Point Condition

$$-8 \sum_{1\leq|\delta|\leq n, 2\leq|\beta|\leq|\alpha|\leq n, i,j,k} \int_0^t \frac{D^\delta \phi_s^i(x) D^\delta \phi_s^j(x) D^\alpha \phi_s^k(x) P_{\beta,x,k}(s)}{(1+\psi_s^{(n)})^2} D_1^i D_2^\beta a^{jk}(x_s, x_s, s) ds$$

$$=: \log(1+d) + I_t + II_t + III_t + IV_t + V_t + VI_t + VII_t + VIII_t + IX_t + X_t. \quad (8.2)$$

Hence for $n \leq 2$ putting $N_t := I_t + II_t$ and using the assumptions on (a, b) we get that there are constants A_n and B_n such that we have the following. We get $III_t + + \ldots + X_t \leq B_n t$ and hence

$$\log\left(1 + \psi_t^{(n)}(x)\right) \leq \log(1+d) + N_t + B_n t \quad (8.3)$$

wherein N_t is a continuous martingale starting in 0 which satisfies $\langle N \rangle_t \leq A_n t$ a.s. for any $t \geq 0$. Thus we get using the Dambis-Dubins-Schwarz' Theorem that there exists a Brownian motion $(W_t)_{t \geq 0}$ such that $\sup_{0 \leq t \leq T} \sum_{1 \leq |\alpha| \leq n, j} |D^\alpha \phi_t^j(x)| \leq (d+1) e^{A_n \sup_{0 \leq t \leq T} W_t + B_n T}$. Using estimates for the geometric Brownian motion as in Chaper 7 shows that for $|\alpha| \leq 2$ we have that

$$\limsup_{T \to \infty} \frac{1}{T} \log \mathbb{P}\left[|D^\alpha \phi_t^j(x)| \geq kT\right] \leq -\frac{(k - B_n)_+^2}{2A_n^2}.$$

To conclude for $|\alpha| > 2$ we make the following observations. The only issue is that the appearance of $P_{\beta,x,j}(s)$ in (8.2) does not allow for the straight conclusion to (8.3). This is overcome in the following way. We only explain the argument because it involves a lot of cumbersome notations and very lengthy formulas that do not make the proof any clearer. Observe that (for some N) the function $f : \mathbb{R}^N \to \mathbb{R}, (x_1, \ldots, x_N) \mapsto \prod_{j=1}^r x_{i_j}$ satisfies for $l, n \in \{1, \ldots, N\}$ that $\partial_l f(x_1, \ldots, x_N) = \sum_{k=1}^r \prod_{k \neq j=1}^r x_{i_j} \delta_{l, i_j}$ and $\partial_n \partial_l f(x_1, \ldots, x_N) = \sum_{k=1}^r \sum_{k \neq m=1}^r \prod_{\{k,m\} \not\ni j=1}^r x_{i_j} \delta_{l,i_j} \delta_{n,i_m}$. Hence if we let enter any finite product of the form $\prod_{i=1}^r D^{\gamma_i} \phi_{0s}^{j_i}(x)$ with $1 \leq |\gamma_i| \leq n$ and $\sum_{i=1}^r |\gamma_i| = \alpha$ squared into the sum in the definition of $\psi_t^{(n)}(x)$ then we may apply $\log(1 + \cdot)$ and end up with (8.3) as before. This completes the proof for general α. □

Remark: In this new approach the dimension d enters into A_n and B_n so we cannot hope to be able to improve the constants obtained for isotropic flows in Chapter 7 in general. The improvement is the much wider scope of the result.

8.3 Proof Of Lemma 8.1.3: The Two-Point Condition

Let us state a general version of Lemma 1.3.8 and 1.4.3.

Lemma 8.3.1 (two-point control general version).
There are constants $\lambda \geq 0$ and $\sigma > 0$ such that for any $x, y \in \mathbb{R}^d$ there is a Brownian motion $(W_t)_{t \geq 0}$ such that we have a.s. that $|x_t - y_t| \leq 2e^{\sigma \sup_{0 \leq s \leq t} W_s + \lambda t}$.

Proof: We assumed more than necessary to apply [52, Lemma 4.1]. □

8.3.1 One-Point Mean Estimates

We will need the following lemma concerning the means of the one-point motion of the spatial derivatives.

Lemma 8.3.2 (mean estimate for the one-point motions).
Assume that $(a, b) \in B^{n,1}_{ub}$. Then there is $\bar{k} > 0$ and $k_n \geq 0$ such that we have for $1 \leq |\alpha| \leq n$, $i \in \{1, \ldots, d\}$, $p \geq 1$ and $x \in \mathbb{R}^d$ that

$$\mathbb{E}\left[\sup_{0 \leq t \leq T} |D^\alpha \phi^i_t(x)|^p\right] \leq \bar{k} e^{k_n T p^2}.$$

Proof: This follows directly from (8.3). □

8.3.2 Two-Point Estimates

Now we come the the proof of Lemma 8.1.3. Let us recall from [20, Proposition 2.3] that for any $T > 0$ there is a constant $C = C(T) > 0$ such that for $x, y \in \mathbb{R}^d$, $j \in \{1, \ldots, d\}$ and $p \geq 1$ we have

$$\mathbb{E}\left[\sup_{0 \leq t \leq T} |D^\alpha \phi^j_t(x) - D^\alpha \phi^j_t(y)|^p\right] \leq e^{Cp^2}(|x - y|^p \wedge 1).$$

Define

$$C_n(T) := \inf_{C>0}\{\forall x, y \in \mathbb{R}^d, 1 \leq |\alpha| \leq n, 1 \leq j \leq d :$$
$$\mathbb{E}\left[\sup_{0 \leq t \leq T} |D^\alpha \phi_t^j(x) - D^\alpha \phi_t^j(y)|^p\right] \leq e^{Cp^2}(|x-y|^p \wedge 1)\}.$$

and observe for $\alpha = \beta + (0,\ldots,0,1,0,\ldots,0)$ (with the 1 at position number l) and $T \leq t \leq 2T$ as before

$$|D^\alpha \phi_t^j(x) - D^\alpha \phi_t^j(y)|^p = |D^l D^\beta \phi_{Tt}^j(x_T) - D^l D^\beta \phi_{Tt}^j(y_T)|^p$$
$$= |\sum_k \partial_l D^\beta \phi_{Tt}^k(x_T) D^k \phi_T^j(x) - \sum_k \partial_l D^\beta \phi_{Tt}^k(y_T) D^k \phi_T^j(y)|^p$$
$$= |\sum_k \partial_l D^\beta \phi_{Tt}^k(x_T)\left[D^k \phi_T^j(x) - D^k \phi_T^j(y)\right] + \sum_k \left[\partial_l D^\beta \phi_{Tt}^k(x_T) - \partial_l D^\beta \phi_{Tt}^k(y_T)\right] D^k \phi_T^j(y)|^p$$
$$\leq (2d)^p \sum_k \sup_{T \leq t \leq 2T} |\partial_l D^\beta \phi_{Tt}^k(x_T)|^p \sup_{0 \leq t \leq T} |D^k \phi_t^j(x) - D^k \phi_t^j(y)|^p$$
$$+ (2d)^p \sum_k \sup_{T \leq t \leq 2T} |\partial_l D^\beta \phi_{Tt}^k(x_T) - \partial_l D^\beta \phi_{Tt}^k(y_T)|^p \sup_{0 \leq t \leq T} |D^k \phi_t^j(y)|^p$$

which yields via integration

$$\mathbb{E}\left[\sup_{0 \leq t \leq 2T} |D^\alpha \phi_t^j(x) - D^\alpha \phi_t^j(y)|^p\right]$$
$$\leq (2d)^p \sum_k \mathbb{E}\left[\sup_{T \leq t \leq 2T} |\partial_l D^\beta \phi_{Tt}^k(x_T)|^p \sup_{0 \leq t \leq T} |D^k \phi_t^j(x) - D^k \phi_t^j(y)|^p\right]$$
$$+ (2d)^p \sum_k \mathbb{E}\left[\sup_{T \leq t \leq 2T} |\partial_l D^\beta \phi_{Tt}^k(x_T) - \partial_l D^\beta \phi_{Tt}^k(y_T)|^p \sup_{0 \leq t \leq T} |D^k \phi_t^j(y)|^p\right]$$
$$= (2d)^p \sum_k \mathbb{E}\left[\mathbb{E}\left[\sup_{T \leq t \leq 2T} |\partial_l D^\beta \phi_{Tt}^k(x_T)|^p \sup_{0 \leq t \leq T} |D^k \phi_t^j(x) - D^k \phi_t^j(y)|^p \middle| \mathcal{F}_T\right]\right]$$
$$+ (2d)^p \sum_k \mathbb{E}\left[\mathbb{E}\left[\sup_{T \leq t \leq 2T} |\partial_l D^\beta \phi_{Tt}^k(x_T) - \partial_l D^\beta \phi_{Tt}^k(y_T)|^p \sup_{0 \leq t \leq T} |D^k \phi_t^j(y)|^p \middle| \mathcal{F}_T\right]\right]$$

$$=(2d)^p \sum_k \mathbb{E}\left[\mathbb{E}\left[\sup_{T\leq t\leq 2T}|\partial_l D^\beta \phi^k_{Tt}(x_T)|^p\Big|\mathcal{F}_T\right]\sup_{0\leq t\leq T}|D^k\phi^j_t(x)-D^k\phi^j_t(y)|^p\right]$$

$$+(2d)^p\sum_k \mathbb{E}\left[\mathbb{E}\left[\sup_{T\leq t\leq 2T}|\partial_l D^\beta \phi^k_{Tt}(x_T)-\partial_l D^\beta \phi^k_{Tt}(y_T)|^p\Big|\mathcal{F}_T\right]\sup_{0\leq t\leq T}|D^k\phi^j_t(y)|^p\right].$$

Applying Lemma 8.3.2 we get that the latter is less or equal to

$$(2d)^p \sum_k \mathbb{E}\left[\bar{k}e^{k_n T p^2}\sup_{0\leq t\leq T}|D^k\phi^j_t(x)-D^k\phi^j_t(y)|^p\right]$$

$$+(2d)^p\sum_k \mathbb{E}\left[e^{C_n(T)p^2}(|x_T-y_T|^p\wedge 1)\sup_{0\leq t\leq T}|D^k\phi^j_t(y)|^p\right]$$

$$\leq (2d)^p d\bar{k}e^{k_n T p^2}e^{C_1(T)p^2}(|x-y|^p\wedge 1)+(2d)^p\sum_k e^{C_n(T)p^2}\mathbb{E}\left[(|x_T-y_T|^p\wedge 1)\sup_{0\leq t\leq T}|D^k\phi^j_t(y)|^p\right]. \tag{8.4}$$

Since we have by Schwarz' inequality and Lemmas 8.3.1, 8.3.2 that

$$\mathbb{E}\left[(|x_T-y_T|^p\wedge 1)\sup_{0\leq t\leq T}|D^k\phi^j_t(y)|^p\right]\leq \sqrt{\mathbb{E}[(|x_T-y_T|^p\wedge 1)^2]\mathbb{E}\left[\sup_{0\leq t\leq T}|D^k\phi^j_t(y)|^{2p}\right]}$$

$$\leq \sqrt{\mathbb{E}[(|x_T-y_T|^{2p})]\wedge 1}\sqrt{\bar{k}e^{4k_n T p^2}}$$

$$\leq ((|x-y|^p e^{\frac{1}{2}(\lambda p+\bar{\sigma}^2 p^2)T}\wedge 1)\sqrt{\bar{k}}e^{2k_n T p^2}$$

$$\leq (|x-y|^p\wedge 1)\sqrt{\bar{k}}e^{\frac{1}{2}(\lambda p+\bar{\sigma}^2 p^2+4k_n p^2)T}.$$

Inserting this into (8.4) we get

$$\mathbb{E}\left[\sup_{0\leq t\leq 2T}|D^\alpha\phi^j_t(x)-D^\alpha\phi^j_t(y)|^p\right]$$

$$\leq 2^p d^{p+1}e^{C_n(T)p^2}\left[\bar{k}e^{k_n T p^2}+\sqrt{\bar{k}}e^{\frac{1}{2}(\lambda p+\bar{\sigma}^2 p^2+4k_n p^2)T}\right](|x-y|^p\wedge 1)$$

$$\leq 2^p d^{p+1}(\bar{k}+\sqrt{\bar{k}})e^{[C_n(T)+\frac{1}{2}(\lambda/p+\bar{\sigma}^2+4k_n)]p^2 T}(|x-y|^p\wedge 1)$$

and hence for $T\geq 1$ and $p\geq 1$ that

$$\frac{C_n(2T)}{2T}\leq 2\frac{\log(2d)+\log(\bar{k}+\sqrt{\bar{k}})}{2T}+\frac{1}{2}\left(\lambda+\bar{\sigma}^2+4k_n\right)+\frac{C_n(T)}{2T}.$$

8.3 Proof Of Lemma 8.1.3: The Two-Point Condition

Assuming w.l.o.g. that $C_n(1) \geq 2\log(2d) + 2\log(\bar{k} + \sqrt{\bar{k}}) + \lambda + \bar{\sigma}^2 + 4k_n$ we get by induction that for any $m \in \mathbb{N}$ we have $\frac{C_n(2^m)}{2^m} \leq \frac{C_n(1)}{1}$. Letting $1 \leq \epsilon \leq 2$ we observe for $m \in \mathbb{N}$ that

$$\frac{C_n(2^m\epsilon)}{2^m\epsilon} \leq \frac{\log(2d) + (\bar{k} + \sqrt{\bar{k}})}{2^{m-1}\epsilon} + \frac{1}{2}\left(\lambda + \bar{\sigma}^2 + 4k_n\right) + \frac{C_n(2^{m-1}\epsilon)}{2^m\epsilon}$$

$$\leq \frac{\log(2d) + (\bar{k} + \sqrt{\bar{k}})}{2^{m-1}} + \frac{1}{2}\left(\lambda + \bar{\sigma}^2 + 4k_n\right) + \frac{C_n(2^m)}{2^m}$$

$$\leq \frac{\log(2d) + (\bar{k} + \sqrt{\bar{k}})}{2^{m-1}} + \frac{1}{2}\left(\lambda + \bar{\sigma}^2 + 4k_n\right) + \frac{C_n(1)}{1}.$$

Hence $\frac{C_n(\cdot)}{(\cdot)}$ is bounded on $[1, \infty)$. Inserting the constant $\bar{c}$ completes the proof because for $s, t \geq 0$ we have $|\log(1+t) - \log(1+s)| \leq |t-s|$ which follows from the mean value theorem. □

Remark: The assumption $(a, b) \in B_{ub}^{n,1}$ enters into the proofs via citation of the results of [20].

Chapter 9

Further Steps To Pesin's Formula

This chapter is devoted to an outlook on further research to be done if one wants to prove Pesin's formula for IOUF's along the lines of [37]. Since assertions concerning uniform continuity cannot be expected to be true in the case of an IOUF we have to replace them by localized versions. If these are not sufficient to yield Pesin's formula, then it is very unlikely that it can be shown in the intended way. Similar considerations apply to uniform Lipschitz continuity etc..

The next thing to do is to spend some time on stable manifolds because they are essential for the partitions constructed to prove the reverse inequality to Theorem 6.3.1. We start with the verification of the following result which is essentially taken from [50]. The continuous-time case has been treated in [39] which we will also use occasionally. Note that using the identification of a stochastic flow and its associated RDS we will use $\varphi_{t-s}(x, \theta^s \omega)$ and $\phi_{s,t}(x,\omega)$ to denote the same object (cf. 1.2.11). In the whole chapter we restrict the attention to the IOUF case since both the IBFs and the RIFs do not posses invariant probabilities.

9.1 Local Stable Manifolds

Theorem 9.1.1 (local stable manifolds for IOUFs).
Let μ be the invariant measure for φ and assume that $\lambda < 0$ is not one of the Lyapunov exponents of φ. Put $\rho := \mathbb{P} \otimes \mu$ which is a probability on $((\text{Diff}(\mathbb{R}^d))^{\mathbb{Z}} \times \mathbb{R}^d, \mathcal{B}(\text{Diff}(\mathbb{R}^d))^{\mathbb{Z}} \otimes \mathcal{B}(\mathbb{R}^d))$ such that $\tau : \Omega \times \mathbb{R}^d \to \Omega \times \mathbb{R}^d$ defined by $\tau((\omega_n)_{n \in \mathbb{Z}}, x) = ((\omega_{n+1})_{n \in \mathbb{Z}}, \varphi_1(x))$ is

ρ-invariant (cf. Definition and Lemma 1.2.4 as well as Lemma 1.2.14). Then there exists a set $\Gamma \subset \Omega \times \mathbb{R}^d$ with $\rho(\Gamma) = 1$ and $\mathcal{B}(\text{Diff}(\mathbb{R}^d))^{\mathbb{Z}} \otimes \mathcal{B}(\mathbb{R}^d)$-measurable real functions $\beta > \alpha > 0$ and $\gamma > 1$ such that we have for $(\omega, z) \in \Gamma$ the following.

1. $\nu^\lambda_{(\omega,z)} := \{x \in K_{\alpha(\omega,z)}(0) : \forall n \geq 0 : \|\varphi_n(z+x,\omega) - \varphi_n(z,\omega)\| \leq \beta(\omega,z)e^{\lambda n}\}$ is a $C^{2,1}$-submanifold of $K_{\alpha(\omega,z)}(0)$ tangent to $\oplus_{\lambda_i < \lambda} E_i(\omega, z)$. Therein $E_i(\omega, z)$ is as in the multiplicative ergodic theorem.

2. If $\lambda' < \lambda$ is chosen such that $[\lambda', \lambda]$ does not contain any Lyapunov exponents of φ then there is $\gamma'(\omega, z)$ such that we have for $x, y \in \nu^\lambda_{(\omega,z)}$ that
$$\|\varphi_n(z+x,\omega) - \varphi_n(z+y,\omega)\| \leq \gamma'(\omega,z) \|x-y\| e^{\lambda' n}.$$

Proof: This is Theorem (5.1) of [50] applied to

- $F : \Omega \times \mathbb{R}^d \to C^{2,1}(\mathbb{R}^d, \mathbb{R}^d), (\omega, z) \mapsto F_{(\omega,z)} = \varphi_1(z+\cdot,\omega) - \varphi_1(z,\omega)$,
- $T_{(\omega,z)} := D_z\varphi_1(\cdot, \omega)$,
- $F^n_{(\omega,z)} := F_{\tau^{n-1}(\omega,z)} \circ \ldots \circ F_{\tau(\omega,z)} \circ F_{(\omega,z)} = \varphi_n(z+\cdot,\omega) - \varphi_n(z,\omega)$ for $n \in \mathbb{N}$,
- $T^n_{(\omega,z)} := T_{\tau^{n-1}(\omega,z)} \circ \ldots \circ T_{\tau(\omega,z)} \circ T_{(\omega,z)} = D_z\varphi_n(\cdot, \omega)$ for $n \in \mathbb{N}$,
- $\tau^n(\omega, z) = (\varphi_n(\omega, z), \theta^n \omega)$.

The identities for $F^n_{(\omega,z)}$, $T^n_{(\omega,z)}$ and $\tau^n(\omega, z)$ are obtained via an easy induction (which we postpone to a more general case in the sequel), so we only have to note that φ is in fact $C^{2,1}$ and that from [57, (2.5)] and [20, (17)] that
$\int_{\mathbb{R}^d \times \Omega} \log^+(\|\varphi_1(z+\cdot,\omega) - \varphi_1(\omega,z)\|_{2,1}) d\rho(\omega,z) < \infty$. □

In fact the proof in [50] also shows the following proposition.

Proposition 9.1.2 (existence of h).
Locally $\nu^\lambda_{(\omega,z)}$ can be written as the graph of a C^2-function from $\oplus_{\lambda_i < \lambda} E_i(\omega, z)$ to $(\oplus_{\lambda_i < \lambda} E_i(\omega, z))^\perp$. In detail this means the following. For any $(\omega, z) \in \Gamma$ there is a function $h_{(\omega,z)} \colon K_{\alpha(\omega,z)}(0) \subset \oplus_{\lambda_i < \lambda} E_i(\omega, z) \to (\oplus_{\lambda_i < \lambda} E_i(\omega, z))^\perp$ with $h_{(\omega,z)}(0) = 0$ such that $\exp_z^{-1} \nu^\lambda_{(\omega,z)}$ is the graph of $h_{(\omega,z)}$.

Proof: There is nothing left to show. The construction of $\nu^\lambda_{(\omega,z)}$ relies on the Arzéla-Ascoli theorem yielding directly the assertion of the above proposition. □

We will need the following versions of stable manifolds. Let $E_0(\omega, z) := \oplus_{\lambda_i < \lambda} E_i(\omega, z)$, $H_0(\omega, z) := E_0(\omega, z)^\perp$ as well as $E_n(\omega, z) := D_z \varphi_n(\cdot, \omega) E_0(\omega, z)$ and $H_n(\omega, z) := D_z \varphi_n(\cdot, \omega) H_0(\omega, z)$. We also need the notations $T_n^l(\omega, z) := D_{\varphi_n(z,\omega)} \varphi_{n+l}(\cdot, \theta^n \omega)$, $S_n^l(\omega, z) := T_n^l(\omega, z)|_{E_n(\omega,z)}$, $U_n^l(\omega, z) := T_n^l(\omega, z)|_{H_n(\omega,z)}$. Now we can formulate the following generalization of Theorem 9.1.1.

Theorem 9.1.3 (more local stable manifolds).
Let μ, λ and ρ be as in Theorem 9.1.1. Then there exists a set $\Gamma \subset \Omega \times \mathbb{R}^d$ with $\rho(\Gamma) = 1$ and for $n \in \mathbb{N}$ there are $\mathcal{B}(\text{Diff}(\mathbb{R}^d))^{\mathbb{Z}} \otimes \mathcal{B}(\mathbb{R}^d)$-measurable real functions $\beta_n > \alpha_n > 0$ and $\gamma_n > 1$ such that we have for $(\omega, z) \in \Gamma$ and $n \in \mathbb{N}$ the following.

1. $\nu^\lambda_{(\omega,z),n} := \{x \in K_{\alpha_n(\omega,z)}(0) : \forall m \geq 0 : \|\varphi_m(\varphi_n(z,\omega) + x, \theta^n \omega) - \varphi_m(\varphi_n(z,\omega), \theta^n \omega)\| \leq \beta_n(\omega, z) e^{\lambda m}\}$ is a $C^{2,1}$-submanifold of $K_{\alpha_n(\omega,z)}(0)$ tangent to $E_n(\omega, z)$.

2. If $\lambda' < \lambda$ is chosen such that $[\lambda', \lambda]$ does not contain any Lyapunov exponents of φ then there is $\gamma'_n(\omega, z) > 1$ such that we have for $x, y \in \nu^\lambda_{(\omega,z),n}$ that
$$\|\varphi_m(\varphi_n(z,\omega) + x, \theta^n \omega) - \varphi_m(\varphi_n(z,\omega) + y, \theta^n \omega)\| \leq \gamma'_n(\omega, z) \|x - y\| e^{\lambda' m}.$$

3. For any $(\omega, z) \in \Gamma$ there is a function $h_{(\omega,z),n} \colon K_{\alpha_n(\omega,z)}(0) \subset E_n(\omega, z) \to H_n(\omega, z)$ with $h_{(\omega,z),n}(0) = 0$ such that $\exp^{-1}_{\varphi_n(z,\omega)} \nu^\lambda_{(\omega,z),n}$ is the graph of $h_{(\omega,z),n}$. Note that the exponential map is of course rather trivial in $\mathbb{R}^d$ but we like to keep the notation close to the manifold case.

4. We have (in fact we may and will choose to have) that
$$\alpha_n(\tau(\omega, z)) \geq \alpha_n(\omega, z) e^\lambda \text{ as well as } \beta_n(\tau(\omega, z)) \geq \beta_n(\omega, z) e^\lambda. \tag{9.1}$$

5. For $x \in \nu^\lambda_{(\omega,z),n}$ we have
$$\varphi_1(\varphi_n(z,\omega) + x, \theta^n \omega) - \varphi_{n+1}(z, \omega) \in \nu^\lambda_{\tau(\omega,z),n+1}. \tag{9.2}$$

6. For $n \in \mathbb{N}$ we have

$$\limsup_{m \to \infty} \frac{1}{m} \log \left(\sup_{x \neq y \in \nu^\lambda_{(z,\omega),n}} \left\{ \frac{|\varphi_m(\varphi_n(z,\omega) + x, \theta^n \omega) - \varphi_m(\varphi_n(z,\omega) + y, \theta^n \omega)|}{|x - y|} \right\} \right) \leq \tilde{\lambda} \quad (9.3)$$

wherein $\tilde{\lambda}$ denotes the largest Lyapunov exponent smaller than λ.

7. If d_ν denotes the distance along $\nu^\lambda_{(\omega,z),m}$ for $m \in \mathbb{N}$ then we have for arbitrary $x, y \in \nu^\lambda_{(\omega,z),n}$ that

$$d_\nu \left(\varphi_m(\varphi_n(z,\omega) + x, \theta^n \omega), \varphi_m(\varphi_n(z,\omega) + y, \theta^n \omega) \right) \leq \gamma'_n(\omega, z) d_\nu(x, y) e^{\lambda m}. \quad (9.4)$$

Remark: If one decreases $\beta_n(\omega, z)$ then one of course obtains the same statement (possibly with a smaller $\alpha_n(\omega, z)$ and a smaller $\gamma_n(\omega, z)$) ending up with a subset of $\nu^\lambda_{(\omega,z),n}$.

Definition 9.1.4 (local stable manifolds).
The set $\nu^\lambda_{(\omega,z),n}$ from Theorem 9.1.3 will be refered to as the nth local stable manifold at z.

Proof of Theorem 9.1.3: The first three assertions are obtained from [50, Theorem (5.1)] applied to

- $F : \Omega \times \mathbb{R}^d \to C^{2,1}(\mathbb{R}^d, \mathbb{R}^d), (\omega, z) \mapsto F_{(\omega,z)} = \varphi_1(\varphi_n(z,\omega) + \cdot, \theta^n \omega) - \varphi_1(\varphi_n(z,\omega), \theta^n \omega)$,

- $T_{(\omega,z)} := D_{\varphi_n(z,\omega)} \varphi_1(\cdot, \theta^n \omega)$,

- $F^m_{(\omega,z)} := F_{\tau_n^{m-1}(\omega,z)} \circ \ldots \circ F_{\tau_n(\omega,z)} \circ F_{(\omega,z)} = \varphi_m(\varphi_n(z,\omega) + \cdot, \theta^n \omega) - \varphi_m(\varphi_n(z,\omega), \theta^n \omega)$
for $m \in \mathbb{N}$,

- $T^m_{(\omega,z)} := T_{\tau_n^{m-1}(\omega,z)} \circ \ldots \circ T_{\tau_n(\omega,z)} \circ T_{(\omega,z)} = D_{\varphi_n(z,\omega)} \varphi_m(\cdot, \theta^n \omega)$ for $m \in \mathbb{N}$,

- $\tau^m(\omega, z) = (\varphi_m(\omega, z), \theta^m \omega)$.

We will now show the identity for $F^m_{(\omega,z)}$. The one for $T^m_{(\omega,z)}$ follows from this as well as the remainder of Theorem 9.1.1 (put $n = 0$ and rename m to be n). The formula for $m = 1$ is

9.1 Local Stable Manifolds

clear by definition so it suffices to observe that

$$F^m_{(\omega,z)}$$
$$= F_{\tau^{m-1}(\omega,z)} \circ F^{m-1}_{(\omega,z)}(x)$$
$$= F_{(\theta^{m-1}\omega, \varphi_{m-1}(z,\omega))} \left(\varphi_{m-1}(\varphi_n(z,\omega) + x, \theta^n\omega) - \varphi_{m-1}(\varphi_n(z,\omega), \theta^n\omega)\right)$$
$$= \varphi_1\left(\varphi_n(\varphi_{m-1}(z,\omega), \theta^{m-1}\omega) + \varphi_{m-1}(\varphi_n(z,\omega) + x, \theta^n\omega) - \varphi_{m-1}(\varphi_n(z,\omega), \theta^n\omega), \theta^{m-1}\theta^n\omega\right)$$
$$\quad - \varphi_1\left(\varphi_n(\varphi_{m-1}(z,\omega), \theta^{m-1}\omega), \theta^n\theta^{m-1}\omega\right)$$
$$= \varphi_m(\varphi_n(z,\omega) + x, \theta^n\omega) - \varphi_m(\varphi_n(z,\omega), \theta^n\omega).$$

The fourth assertion is now seen to be [39, (3.29)]. The proof of (9.2) is as follows. Since we wrote $\exp^{-1}_{\varphi_n(z,\omega)} \nu^\lambda_{(\omega,n),z}$ as the graph of $h_{(\omega,z),n}$ we get

$$\left\{(\xi, h_{(\omega,z),n}(\xi)): \xi \in K_{\alpha_n(\omega,z)}(0) \cap E_n(\omega,z)\right\}$$
$$= \exp^{-1}_{\varphi_n(z,\omega)} \nu^\lambda_{(\omega,n),z}$$
$$= \exp^{-1}_{\varphi_n(z,\omega)} \left\{x \in K_{\alpha(\omega,z)}(\varphi_n(z,\omega)): \forall m \geq 0: \right.$$
$$\left. \|\varphi_m(\varphi_n(z,\omega) + x, \theta^n\omega) - \varphi_m(\varphi_n(z,\omega), \theta^n\omega)\| \leq \beta_n(\omega,z)e^{\lambda m}\right\}$$
$$= \left\{y \in K_{\alpha(\omega,z)}(0): \forall m \geq 0: \|\varphi_m(y, \theta^n\omega) - \varphi_m(\varphi_n(z,\omega), \theta^n\omega)\| \leq \beta_n(\omega,z)e^{\lambda m}\right\}.$$

This by the way proves that for any $\xi \in K_{\alpha_n(\omega,z)}(0) \cap E_n(\omega,z)$ there is a unique $\eta \in E_n(\omega,z)^\perp$ such that for all $m \in \mathbb{N}$ we have

$$\|\varphi_m(\varphi_n(z+\xi+\eta),\omega), \theta^n\omega) - \varphi_m(\varphi_n(z,\omega), \theta^n\omega)\| \leq \beta_n(\omega,z)e^{\lambda m}.$$

This is an analogue to [37, III.(3.3)]. Observe now for $x \in \nu^\lambda_{(\omega,z),n}$ that

$$\left\|\varphi_m(\varphi_{n+1}(z,\omega) + \varphi_1(\varphi_n(z,\omega) + x, \theta^n\omega) - \varphi_{n+1}(z,\omega), \theta^{n+1}\omega) - \varphi_m(\varphi_{n+1}(z,\omega), \theta^{n+1}\omega)\right\|$$
$$= \left\|\varphi_m\left(\varphi_1(\varphi_n(z,\omega) + x, \theta^n\omega), \theta^{n+1}\omega\right) - \varphi_m(\varphi_n(\varphi_1(z,\omega), \theta\omega), \theta^n\omega)\right\|$$
$$= \left\|\varphi_m\left(\varphi_1(\varphi_n(z,\omega) + x, \theta^n\omega), \theta^{n+1}\omega\right) - \varphi_m(\varphi_1(\varphi_n(z,\omega), \theta^n\omega), \theta^{n+1}\omega)\right\|$$
$$= \left\|\varphi_{m+1}\left(\varphi_n(z,\omega) + x, \theta^n\omega\right) - \varphi_{m+1}\left(\varphi_n(z,\omega), \theta^n\omega\right)\right\|$$
$$\leq \beta_n(\omega,z)e^{\lambda(m+1)} \leq \beta_n(\tau(\omega,z))e^{\lambda m}$$

by (9.1). This proofs (9.2). (9.3) is a direct consequence of [39, Theorem 3.1 (b)]. (9.4) now follows from this and the mean value theorem via integration. □

9.2 Global Stable Manifolds

The local stable manifolds $\nu^\lambda_{(\omega,z),n}$ consist of points in the neighbourhood of $z \in \mathbb{R}^d$ approaching the trajectory of z exponentially fast. The set of points with the latter property is seen to be a manifold which we call the nth global stable manifold of z. It of course is a random manifold.

Definition 9.2.1 (global stable manifolds).
Let λ be as in Theorems 9.1.1 and 9.1.3. The set

$$\nu^{\lambda,g}_{(\omega,z),n} := \{x \in \mathbb{R}^d \colon \lim_{m \to \infty} \frac{1}{m} \log \|\varphi_m(\varphi_n(z,\omega) + x, \theta^n \omega) - \varphi_m(\varphi_n(z,\omega), \theta^n \omega)\| \leq \lambda\}$$

is called the nth global stable manifold of z. The 0th global stable manifold of z will just be refered to as the global stable manifold of z.

We have the following theorem about the global stable manifolds.

Theorem 9.2.2 (properties of global stable manifolds).
Let $\nu^{\lambda,g}_{(\omega,z),n}$ be as in Definition 9.2.1 and $\nu^\lambda_{(\omega,z),n}$ be as in Theorem 9.1.3. Then the following holds true.

1. *For any $m \in \mathbb{N}$ we have $(\varphi_m(\varphi_n(z,\omega) + \cdot, \theta^n \omega) - \varphi_m(\varphi_n(z,\omega), \theta^n \omega)) \nu^{\lambda,g}_{(\omega,z),n} = \nu^{\lambda,g}_{(\omega,z),n+m}$*

2. *Suppose we are working in the natural extension of our RDS (which we do without change of notation by just assuming that $\varphi_k(\cdot,\cdot)$ makes sense for negative $k \in \mathbb{Z}$). Then defining the increasing family of random sets $(\nu^{\lambda,m}_{(\omega,z),n})_{m \in \mathbb{N}}$ as $\nu^{\lambda,0}_{(\omega,z),n} = \nu^\lambda_{\tau^m(\omega,z),n}$ and*

    $\nu^{\lambda,m}_{(\omega,z),n} := \left(\varphi_{-m}(z + \cdot, \theta^{m+n}\omega) \nu^\lambda_{(\theta^m \omega,z),n} - \varphi_{-m}(z, \theta^{m+n}\omega)\right) \nu^\lambda_{(\theta^m \omega,z),n}$ *if*
    $\nu^{\lambda,m-1}_{(\omega,z),n} \subset \left(\varphi_{-m}(z + \cdot, \theta^{m+n}\omega) \nu^\lambda_{(\theta^m \omega,z),n} - \varphi_{-m}(z, \theta^{m+n}\omega)\right)$ *and $\nu^{\lambda,m}_{(\omega,z),n} := \nu^{\lambda,m-1}_{(\omega,z),n}$ otherwise*

we have $\mathbb{P} \otimes \rho$-a.s. that
$$\nu^{\lambda,g}_{(\omega,z),n} = \bigcup_{m=0}^{\infty} \nu^{\lambda,m}_{(\omega,z),n}. \tag{9.5}$$

Proof: Despite the fact that similar proofs can be found in [12] and [39] we give some details. The first assertion can be shown in the following manner.

$$(\varphi_m(\varphi_n(z,\omega) + \cdot, \theta^n \omega) - \varphi_{n+m}(z,\omega)) \nu^{\lambda,g}_{(\omega,z),n}$$
$$= \{\varphi_m(\varphi_n(z,\omega) + x, \theta^n \omega) - \varphi_{n+m}(z,\omega):$$
$$\limsup_{k\to\infty} \frac{1}{k} \log \|\varphi_k(\varphi_n(z,\omega) + x, \theta^n \omega) - \varphi_{n+k}(z,\omega)\| \leq \lambda\}$$
$$= \left\{y: \limsup_{k\to\infty} \frac{1}{k} \log \|\varphi_k(\varphi_n(z,\omega)\right.$$
$$\left. + \varphi_m^{-1}(y + \varphi_{m+n}(z,\omega), \theta^n \omega) - \varphi_n(z,\omega)), \theta^n \omega) - \varphi_{n+k}(z,\omega)\| \leq \lambda\right\}$$
$$= \left\{y: \limsup_{k\to\infty} \frac{1}{k} \log \|\varphi_k(\varphi_m^{-1}(y + \varphi_{m+n}(z,\omega), \theta^n \omega), \theta^n \omega) - \varphi_{n+k}(z,\omega)\| \leq \lambda\right\}$$
$$= \left\{y: \limsup_{k\to\infty} \frac{1}{k} \log \|\varphi_k(\varphi_{-m}(y + \varphi_{m+n}(z,\omega), \theta^{n+m} \omega), \theta^n \omega) - \varphi_{n+k}(z,\omega)\| \leq \lambda\right\}$$
$$= \left\{y: \limsup_{k\to\infty} \frac{1}{k} \log \|\varphi_{k-m}(y + \varphi_{m+n}(z,\omega), \theta^{n+m} \omega) - \varphi_{n+k}(z,\omega)\| \leq \lambda\right\}$$
$$= \left\{y: \limsup_{k\to\infty} \frac{1}{k+m} \log \|\varphi_k(y + \varphi_{m+n}(z,\omega), \theta^{n+m} \omega) - \varphi_{n+m+k}(z,\omega)\| \leq \lambda\right\}$$
$$= \left\{y: \limsup_{k\to\infty} \frac{1}{k} \log \|\varphi_k(y + \varphi_{m+n}(z,\omega), \theta^{n+m} \omega) - \varphi_{n+m+k}(z,\omega)\| \leq \lambda\right\}$$
$$= \nu^{\lambda,g}_{(\omega,z),n+m}.$$

The proof of (9.5) is similar to the proof of [39, (3.38)] and we omit the details. See also [12, Lemma 3.2.2] for the case $z = 0$ is a fixed point. $\square$

9.3 Local Hölder Continuity Of Oseledec' Splittings

We now start to discuss the dependence of $E_0(\omega, z)$ and $H_0(\omega, z)$ on $z \in \mathbb{R}^d$. First we recall some facts from [37] which are stated in the compact manifold setting but with proofs that perfectly cover our case. In this whole subsection let λ be as before and let $[\lambda, \bar{\lambda}] \subset$ (-

$\infty, 0]$ be a compact interval that does not contain any Lyapunov exponents of φ. We fix $0 < \epsilon \leq (\bar{\lambda} - \lambda)/200$.

Lemma 9.3.1 (definition of $l(n, \omega, x)$).
There exists a measurable function $l : \Gamma \times \mathbb{N} \to (0, \infty)$ such that we have for $(\omega, z) \in \Gamma$ and $n, l \in \mathbb{N}$ the following.

$$\forall \xi \in E_n(\omega, z) \colon |S_n^l(\omega, z)\xi| \leq l(\omega, z, n) e^{(\lambda+\epsilon)l} |\xi|,$$

$$\forall \eta \in H_n(\omega, z) \colon |U_n^l(\omega, z)\eta| \geq l(\omega, z, n)^{-1} e^{(\bar{\lambda}-\epsilon)l} |\xi|,$$

$$\gamma(E_{n+l}(\omega, z), H_{n+l}(\omega, z)) \geq l(\omega, z, n)^{-1} e^{-\epsilon l}$$

wherein γ denotes the angle between two vector spaces.

$$l(\omega, z, n+l) \leq l(\omega, z, n) e^{\epsilon l}$$

Proof: Since an IOUF is ergodic this is just [37, Lemma III.1.1] because the Lyapunov spectrum is constant. □

Let $l' \in \mathbb{R}$ be a number such that the set $\Gamma_\epsilon^{l'} := \{(\omega, z) \in \Gamma \colon l(\omega, z, 0) \leq l'\} \neq \emptyset$. We then have the following lemma concerning the continuity of $E_0(\omega, z)$ and $H_0(\omega, z)$.

Lemma 9.3.2 (continuity of $E_0(\omega, z)$).
$E_0(\omega, z)$ and $H_0(\omega, z)$ depend continuously on $(\omega, z) \in \Gamma_\epsilon^{l'}$.

Proof: Although this is [37, Lemma III.1.2] we have to comment on its meaning. The topology on the Grassmanian manifold consisting of k-dimensional subspaces of $\mathbb{R}^d$ can be assumed to be canonical (in special cases it is just a suitable projective space) and the topology on $\mathbb{R}^d$ is of course clear, but it is not really common to have the probability space equipped with a topological structure to study the continuity of random variables. The topology that is used on Ω in [37] is the countable product of the C^2-topology which is not suitable for IOUFs since their C^2-norms are a.s. infinite. The assertion of the above lemma may nevertheless be used as the statement that for almost all $\omega \in \Omega$ the mapping $\mathbb{R}^d \ni z \mapsto E_0(\omega, z)$ or $H_0(\omega, z)$ respectively is continuous. A serious inspection of the proof also shows the following. The above convergence i.e. the proof of [37] also holds if one replaces the C^2-topology on Diff($\mathbb{R}^d$) by the topology of uniform C^2-convergence on compacts. This

9.3 Local Hölder Continuity Of Oseledec' Splittings

can be seen in the following way. Let $(\omega, x)_n \to (\omega, x)$. Using the compactness of the Grassmannian manifold we get that $E_0((\omega, x)_n)$ contains a convergent subsequence with a limit E (we pass to the subsequence without changing notation). Since manifolds are metric spaces it is sufficient to show that $E = E_0(\omega, x)$. In view of Lemma 9.3.1 it is sufficient to show that for any $l \in \mathbb{N}$ and $\xi \in E$ we have

$$|S_0^l(\omega, x)\xi| \leq Ke^{(\lambda+\epsilon)l}$$

for a suitable constant K (independent of l). To see this it is sufficient to observe for any sequence $E_0((\omega, x)_n) \ni \xi_n \to \xi \in E$ that

$$|S_0^l(\omega, x)\xi| \leq |S_0^l(\omega, x)(\xi - \xi_n)| + |(S_0^l(\omega, x) - S_0^l((\omega, x)_n))\xi_n| + |S_0^l((\omega, x)_n)\xi_n|$$
$$\leq \left\|S_0^l(\omega, x)\right\| |\xi - \xi_n| + \left\|(S_0^l(\omega, x) - S_0^l((\omega, x)_n))\right\| |\xi_n| + l(\omega_n, x_n, 0)e^{(\lambda+\epsilon)l}|\xi_n|$$
$$\leq \left\|S_0^l(\omega, x)\right\| |\xi - \xi_n| + \left\|(S_0^l(\omega, x) - S_0^l((\omega, x)_n))\right\| |\xi_n| + l'e^{(\lambda+\epsilon)l}|\xi_n|.$$

Since $\{\xi_n : n \in N\}$ is compact we have that the first two terms vanish as $n \to \infty$ and the proof is complete. □

The following proposition will be used to show even a bit more than continuity of $E_0(\omega, x)$ and $H_0(\omega, x)$. We first have to define the notion of Hölder continuity for subspaces of $\mathbb{R}^d$.

Definition 9.3.3 (local Hölder continuity).
Let X be a metric space and $(E_x : x \in X)$ be a family of subspaces of the Hilbert space H. The family $(E_x : x \in X)$ is said to be Hölder continuous with exponent $\alpha > 0$ and constant $C > 0$ if we have for any $x, y \in X$ that $d(E_x, E_y) := \mathrm{Ap}(E_x, E_y) \vee \mathrm{Ap}(E_y, E_x) \leq Cd(x, y)^\alpha$. Therein $\mathrm{Ap}(E, \tilde{E}) := \sup_{\xi \in E, \|\xi\|=1} \inf_{\eta \in \tilde{E}} \|\xi - \eta\|$ is the aperture between E and $\tilde{E}$.

Proposition 9.3.4 (sufficient conditions for Hölder continuity).
Let (X, d) be a metric space with diameter at most 1. Let H be a Hilbert space and $(T_i(x) : x \in X, i \in \mathbb{N})$ be a family of bounded linear operators $T_i(x) \colon H \to H$. Write $T^0(x) = \mathrm{id}$ and $T^n(x) := T_n(x) \circ \ldots \circ T_1(x)$. For numbers $\hat{C} \geq 1$, $\hat{a} < \hat{b}$ let $\Delta_{\hat{a}, \hat{b}, \hat{C}}$ the (maybe empty) set of points $x \in X$ for which there exist splittings $H = E_x \oplus E_x^\perp$ such that for any $n \in \mathbb{N}$ the following holds.

$$\|T^n(x)\xi\| \leq \hat{C}e^{\hat{a}n}\|\xi\|, \xi \in E_x \text{ and } \|T^n(x)\eta\| \geq \hat{C}^{-1}e^{\hat{b}n}\|\eta\|, \xi \in E_x^\perp.$$

Suppose there are numbers $\hat{c} > \hat{a}$ and $\beta > 0$ such that for $x, y \in X$ and $n \in \mathbb{N}$ we have

$$||T^n(x) - T^n(y)|| \leq e^{\hat{c}n} d(x,y)^\beta.$$

Then the family $(E_x : x \in \Delta_{\hat{a},\hat{b},\hat{C}})$ is Hölder continuous with exponent $\frac{\hat{a}-\hat{b}}{\hat{a}-\hat{c}}\beta$ and constant $3\hat{C}^3 e^{\hat{b}-\hat{a}}$.

Proof: [37, Proposition III.4.1]. □

The application of Proposition 9.3.4 now yields the following theorem.

Theorem 9.3.5 (local Hölder continuity).
Then the family $(E_0(\omega, x) : (\omega, x) \in \Gamma_\epsilon^{l'})$ is a.s. locally Hölder continuous w.r.t. $x \in \mathbb{R}^d$. This means in detail for $(\omega, x) \in \Gamma_\epsilon^{l'}$ that there exists $\alpha \geq \lambda + \epsilon$ such that the family $\left(E_0(\omega, y) : y \in K_1(x) \cap (\Gamma_\epsilon^{l'})_\omega\right)$ is Hölder-continuous with constant $3l'^2 e^{\bar{\lambda} - \lambda - 2\epsilon}$ and exponent $\frac{\bar{\lambda}-\lambda-2\epsilon}{\alpha-\lambda-\epsilon}$.

Proof: The proof follows from Lemma 9.3.1 and Proposition 9.3.4 applied to $H = \mathbb{R}^d$, $X = K_1(x)$, $E_x = E_0(\omega, x)$, $T_n(x) = D_{\varphi_{n-1}(x,\omega)} \varphi_1(x, \theta^{n-1}\omega)$ and $\Delta_{\hat{a},\hat{b},\hat{C}} = \Gamma_\epsilon^{l'}$ since we have the following. For all $x \in \mathbb{R}^d$ there is an a.s. finite random number $\alpha = \alpha(\omega)$ such that for all $n \in \mathbb{N}$ and $|x-y| \leq 1$ we have

$$||D_x \phi_n(\cdot) - D_y \phi_n(\cdot)|| \leq e^{\alpha n} |x - y|.$$

This follows from Theorem 8.1.1 since the local characteristics of IOUFs allow for the application to the second order derivatives. □

Remark: To further proceed with the proof of Pesin's formula along the lines of [37] one might wish to start at [37, p. 84, III.5].

Chapter 10

Some Open Problems

In this chapter we name some questions as possible directions of further research. We order them by the chapter they arise although there is almost no reference to these questions in the text before. We conjecture that the difficulty of these problems varies from an advanced graduate exercise level to a quite difficult level, that cannot be captured without a significant amount of work.

10.1 Open Problems arising from Chapter 2

In Chapter 2 we gave a spatial regularity lemma which suggests to ask for the following refinements.

Problem 10.1.1 (precise spatial asymptotics).
Given an IBF, IOUF or a RIF with convariance tensor b and drift c ($c = 0$ for the IBF case) find the right order of magnitude of $\sup_{|x|\geq R}(|\phi_t(x) - e^{-ct}x|)$ i.e. find a function f such that
$$\lim_{R\to\infty} \sup_{|x|\geq R} \frac{|\phi_t(x) - e^{-ct}x|}{f(|x|)} \text{ exists in a non-trivial way.}$$
Therein „non-trivial" means that the limit is strictly positive and finite w.p.p..

Problem 10.1.2 (precise spatial asymptotics 2).
Find a solution to Problem 10.1.1 that is valid for a larger class of stochstic flows or at least prove Lemma 2.2.1 in this context.

These problems seem to be rather feasible but we confined ourselves to the treatment given in the text because it is sufficient for the application in Chapter 5.

10.2 Open Problems arising from Chapter 3

The question of positivity of densities for the n-point motion seems to be still not completely solved. It is possible that quite standard methods lead to a solution to the following problems.

Problem 10.2.1 (positivity of densities).
Find more reasonable conditions on the covariance tensor b such that the two-point motion of the associated IBF (or IOUF or RIF) possesses a strictly positive density on $\mathbb{R}^d_\times$.

Problem 10.2.2 (positivity of densities 2).
Solve Problem 10.2.1 for the n-point motion with the obvious analogue for $\mathbb{R}^d_\times$. Observe that it is necessary to ensure the existence of a density by investigation of the hypoellipticity of the diffusion matrix.

Since both these problems do not play a central role in the later chapters we did not pay attention to them.

10.3 Open Problems arising from Chapter 4

The following problems seem to be quite hard to tackle.

Problem 10.3.1 (a.s. convergence in the limit shape theorem).
Prove (or disprove) that in the limit shape theorem one has a.s. convergence instead of convergence in probability yielding that the limit shape property holds for all sufficiently large times T.

Problem 10.3.2 (general limit shape theorem).
Find the limiting behaviour as in the limit shape theorem for IOUFs (with sub-linear scaling) or RIF (with exponential scaling).

Remark: In both cases we conjecture the limiting behaviour to be random.

Problem 10.3.3 (general limit shape theorem 2).
Find the limiting behavior in higher dimension or for a more general class of stochastic flows. Does the expansion speed depend on the codimension of the initial set? Is it still well-defined in some sense?

10.4   Open Problems arising from Chapter 5

We indicated that it is not straightforward to find a Lyapunov cohomology between the RDS coming from an IOUF and a C^2-RDS on the $d+1$-dimensional unit ball and we strongly conjecture that it is impossible to find one. Nevertheless one might ask the following.

Problem 10.4.1 (Lyapunov cohomology).
Prove that there is no Lyapunov cohomology between the RDS coming from an IOUF and any C^2-RDS living on a compact Riemannian manifold or find one.

10.5   Open Problems arising from Chapter 6 and Chapter 9

The following problems directly arise from the fact that we did not need so much IOUF structure in Chapter 6.

Problem 10.5.1 (Ruelle's inequality).
Generalize Theorem 6.3.1 to the case of a general stochastic flow with one-point motions that have invariant probabilities.

Problem 10.5.2 (Ruelle's inequality).
Find a reasonable notion of entropy for IBFs (or even RIFs) and generalize Theorem 6.3.1 accordingly.

Problem 10.5.3 (Pesin's formula).
Prove equality in Theorem 6.3.1 or even in one of the settings of Problems 10.5.1 or 10.5.2.

10.6 Open Problems arising from Chapter 7 and Chapter 8

Except from the improvement of the constants given as upper bounds there is just one problem left

Problem 10.6.1 (lower bound).
Find a lower bound for the exponential expansion rate of the derivative in any of the treated cases of stochastic flows.

Bibliography

[1] Ludwig Arnold. *Random dynamical systems*. Springer Monographs in Mathematics. Springer, Berlin, 1998.

[2] Ludwig Arnold and Michael Scheutzow. Perfect cocycles through stochastic differential equations. *Probab. Theory Relat. Fields*, 101(1):65–88, 1995.

[3] Jörg Bahnmüller and Thomas Bogenschütz. A Margulis-Ruelle inequality for random dynamical systems. *Arch. Math.*, 64:246–253, 1995.

[4] Luis Barreira and Yakow Pesin. *Nonuniform Hyperbolicity: Dynamics Of Systems With Nonzero Lyapunov Exponents*. Cambridge University Press, 2007.

[5] Peter Baxendale and Theodore E. Harris. Isotropic stochastic flows. *Ann. Probab.*, 14(2):1155–1179, 1986.

[6] Denis Bell. *The Malliavin Calculus*. Dover Publications, Inc, 2006.

[7] Jean-Michel Bismut. *Large Deviations And The Malliavin Calculus*, volume 45 of *Progress In Mathematics*. Birkhäuser, Boston, 1984.

[8] Vladimir I. Bogachev. *Gaussian measures*, volume 62 of *Mathematical Surveys and Monographs*. American Mathematical Society, Providence, RI, 1998.

[9] Michael Cranston, Michael Scheutzow, and David Steinsaltz. Linear expansion of isotropic Brownian flows. *Elect. Comm. in Probab.*, 4:91–101, 1999.

[10] Michael Cranston, Michael Scheutzow, and David Steinsaltz. Linear bounds for stochastic dispersion. *Ann. Probab.*, 28(4):1852–1869, 2000.

[11] Burgess Davis. On the L^p-norms of stochastic integrals and other martingales. *Duke Mathematical Journal*, 43(4), 1976.

[12] Georgi Dimitroff. *Some properties of isotropic Brownian and Ornstein-Uhlenbeck flows.* PhD thesis, TU Berlin, 2006. URL: http://opus.kobv.de/tuberlin/volltexte/2006/1252/.

[13] Dmitry Dolgopyat, Vadim Kaloshin, and Leonid Koralov. A limit shape theorem for periodic stochastic dispersion. *Comm. Pure Appl. Math.*, 57(9):1127–1158, 2004.

[14] Ilja N. Bronstein et al. *Taschenbuch der Mathematik.* Verlag Harri Deutsch, 2001.

[15] Albert Fathi, Michel Herman, and Jean-Christophe Yoccoz. *A proof of Pesin's stable manifold theorem*, volume 1007. Springer, 1983.

[16] Avner Friedman and Mark Pinsky. *Stochastic Analysis.* Academic Press, 1978.

[17] Theodore E. Harris. Brownian motions on the homeomorphisms of the plane. *The Annals of Probability*, 9(2):232–254, 1981.

[18] Lars Hörmander. Hypoelliptic second order differential equations. *Acta Math.*, 119:147–171, 1969.

[19] Nobuyuki Ikeda and Shinzo Watanabe. *Stochastic differential equations and diffusion processes*, volume 24 of *North-Holland mathematical library*. North-Holland publ. comp., 1981.

[20] Peter Imkeller and Michael Scheutzow. On the spatial asymptotic behaviour of stochastic flows in Euklidean space. *Ann. Probab.*, 27:109–129, 1999.

[21] Ioannis Karatzas and Steven Shreve. *Brownian Motion And Stochastic Calculus Second Edition.* Springer, 1991.

[22] Anatole Katok and Jean-Marie Strelcyn. *Invariant manifolds, entropy and billiards; smooth maps with singularities*, volume 1222 of *Lecture Notes In Mathematics*. Springer, 1986.

[23] Gerhard Keller. *Equilibrium states in ergodic theory.* London mathematical society. Cambridge University Press, 1998.

[24] Yuri Kifer. *Ergodic theory of random transformations*, volume 10 of *Progress In Probability And Statistics*. Birkhäuser, Cambridge, UK, 1996.

[25] Sandra Kliem. Bedingte stochastische Flüsse und ihre Anwendung auf die Abschätzung des Durchmesser des Bildes einer Menge unter einem solchen Fluss. Diploma Thesis TU Berlin, 2004.

[26] Konrad Königsberger. *Analysis 1*. Springer, fifth edition, 2001.

[27] Hiroshi Kunita. *Stochastic Flows and Stochastic Differential Equations*. Cambridge University Press, Cambridge, UK, 1990.

[28] Yves Le Jan. On isotropic Brownian motions. *Z. Wahrscheinlichkeitstheor. Verw. Geb.*, 70:609–620, 1985.

[29] Remi Leandre. Positivity theorem for a general manifold. *SORT*, 29(1):11–26, 2005.

[30] François Ledrappier. *Quelques propriétés des exposants charactéristiques*, volume 1097 of *Lecture Notes In Mathematics*. R. M. Dudley, H. Kunita, F. Ledrappier: École d'Été de Probabilité de Saint-Flour XII -1982, 1982.

[31] François Ledrappier and Lai-Sang Young. The metric entropy of diffeomorphisms. *Bull. Amer. Math. Soc. (N.S.)*, 11(2):343–346, 1984.

[32] François Ledrappier and Lai-Sang Young. The metric entropy of diffeomorphisms. I. Characterization of measures satisfying Pesin's entropy formula. *Ann. of Math. (2)*, 122(3):509–539, 1985.

[33] François Ledrappier and Lai-Sang Young. The metric entropy of diffeomorphisms. II. Relations between entropy, exponents and dimension. *Ann. of Math. (2)*, 122(3):540–574, 1985.

[34] François Ledrappier and Lai-Sang Young. Dimension formula for random transformations. *Comm. Math. Phys.*, 117(4):529–548, 1988.

[35] Hannelore Lisei and Michael Scheutzow. Linear bounds and Gaussian tails in a stochastic dispersion model. *Stochastics and Dynamics*, 1(3):389–403, 2001.

[36] Hannelore Lisei and Michael Scheutzow. On the dispersion of sets under the action of an isotropic Brownian flow. In *Probabilistic Methods in Fluids*, pages 224–238. World Scientific, Singapore, 2003.

[37] Pei-Dong Liu and Min Qian. *Smooth ergodic theory of random dynamical systems*, volume 1606 of *Lecture Notes In Mathematics*. Springer, 1995.

[38] Paul Malliavin. Stochastic claculus of variations and hypoelliptic operators. *proc. of the intern. conf. on SDE, Kyoto*, pages 195–263, 1976.

[39] Salah-Eldin A. Mohammed and Michael K.R. Scheutzow. The stable manifold theorem for stochastic differential equations. *Ann. Probab.*, 27(2):615–652, 1999.

[40] James Norris. Simplified Mallivin calculus. *Seminaire de probabilité XIX*, 1985.

[41] David Nualart. *The Malliavin calculus and related topics*. Springer, 1995.

[42] Taijiro Ohno. Asymptotic behaviours of dynamical systems with random parameters. *Publ. R.I.M.S. Kyoto University*, 19:83–98, 1983.

[43] Yakov Pesin. Families of invariant manifolds corresponding to non-zero characteristic exponents. *Math of the USSR Izvestija*, 10(6):1261–1305, 1976.

[44] Yakov Pesin. Lyapunov characteristic exponents and smooth ergodic theory. *Russ Math. Surveys*, 32(4):55–114, 1977.

[45] Leonid Piterbarg. Relative dispersion in 2d stochastic flows. *Journal of Turbulence*, 6:1–18, 2005.

[46] Daniel Revuz and Marc Yor. *Continuous martingales and Brownian motion*, volume 293 of *Grundlehren der Mathematischen Wissenschaften [Fundamental Principles of Mathematical Sciences]*. Springer-Verlag, Berlin, third edition, 1999.

[47] Paolo Emilio Ricci. Improving the asymptotics for the greatest zeros of hermite polynomials. *Computers Math. Appl.*, 30:409–416, 1999.

[48] Vladimir Abramovich Rokhlin. Lectures on the entropy theory of measure preserving transformations. *Russ. Math. Surv.*, 22(5):1–52, 1967.

[49] David Ruelle. An inequality for the entropy of differential maps. *Bol. Soc. Bras. de Math.*, 9:83–87, 1978.

[50] David Ruelle. Ergodic theory of differentiable dynamical systems. *Publ. Math IHES*, 50:27–58, 1979.

[51] Giovanni Sansone and Roberto Conti. *Non-Linear Differential Equations*. Pergamon Press, 1964.

[52] Michael Scheutzow. Chaining techniques and their application to stochastic flows. *Trends in Stochastic Analysis*, 353:35–63, 2009.

[53] Michael Scheutzow and David Steinsaltz. Chasing balls through martingale fields. *Ann. Probab.*, 30(4):2046–2080, 2002.

[54] Holger van Bargen. Asymptotic growth of spatial derivatives of isotropic flows. *Electronic Journal of Probability*, 14:2328–2351, 2009.

[55] Holger van Bargen. Ruelle's inequality for isotropic Ornstein-Uhlenbeck flows. *Stochastics and Dynamics*, 10(1):143–154, 2009. http://dx.doi.org/10.1142/S0219493710002802.

[56] Holger van Bargen. A weak limit shape theorem for planar isotropic Brownian flows. *preprint*, 2009.

[57] Holger van Bargen and Georgi Dimitroff. Isotropic Ornstein-Uhlenbeck flows. *Stochastic Processes And Applications*, 119 (7)(1):2166–2197, 2009.

[58] Peter Walters. *Ergodic theory - introductory lectures*, volume 458 of *Lecture Notes In Mathematics*. Springer, 1975.

[59] Shinzo Watanabe. Analysis of Wiener functionals (Mallivin calculus) and its application to heat kernels. *The Annals Of Probability*, 15(1):1–39, 1987.

[60] Akiva Moiseevich Yaglom. Some classes of random fields in n-dimensional space, related to stationary random processes. *Theory of Probability and its Applications*, 28:273–320, 1957.

[61] Lai Sang Young. Dimension, entropy and Lyapunov exponents. *Ergodic Theory Dynamical Systems*, 2(1):109–124, 1982.

Index Of Notation and Abbreviations

$(\cdot)^+$	$:= (\cdot) \vee 0$ - the positive part of $(\cdot)$
$\|\cdot\|$	Euclidean norm
$\|f\|_{m:\mathbb{K}}$	$:= \sup_{x\in\mathbb{K}} \frac{|f(x)|}{1+|x|} + \sum_{1\le|\alpha|\le m} \sup_{x\in K} |\mathrm{D}_x^\alpha f(x)|$
$\|g\|_{m:\mathbb{K}}^{\sim}$	$:= \sup_{x\in K} \frac{|g(x,y)|}{(1+|x|)(1+|y|)} + \sum_{1\le|\alpha|\le m} \sup_{x,y\in\mathbb{K}} |\mathrm{D}_y^\alpha \mathrm{D}_x^\alpha g(x,y)|$
$\|g\|_{m+\delta:\mathbb{K}}^{\sim}$	$:= \|g\|_{m:\mathbb{K}}^{\sim} + \sum_{|\alpha|=m} \|\mathrm{D}_x^\alpha \mathrm{D}_y^\alpha g\|_{\delta:\mathbb{K}}^{\sim}$
$\|g\|_{m+\delta}^{\sim}$	$:= \|g\|_{m+\delta:\mathbb{D}}^{\sim}$
$\langle\cdot\rangle, \langle\cdot,\cdot\rangle$	cross variation bracket
$\|\cdot\|_\infty$	supremum norm
$\|\cdot\|$	operator norm coming from $|\cdot|$
∇	the nabla operator
$\overline{x,y}$	convex hull of x and y
$I, II, III, \ldots, I_t, II_t, \ldots$	terms to be estimated
$|v|^R$	$:= \sup_{\gamma\in\mathcal{C}_R} \mathbb{E}\left[\tau^R(\gamma,v)\right]$
$\|v\|^R$	$:= \lim_{t\to\infty} \frac{|tv|^R + C_{12}}{t}$ - stable norm
$\|v\|_{\tilde{R}}^R$	$:= \lim_{t\to\infty} \frac{\sup_{\gamma\in\mathcal{C}_{\tilde{R}}} \mathbb{E}[\tau^R(\gamma,tv)]}{t}$
$\subset\subset$	relatively compact subset
$a(x,y,s) = (a^{ij}(x,y,s))_{1\le i,j\le n}$	cross variation part of local characteristic
A	element of $\in \mathcal{B}(C(\mathbb{R}^d,\mathbb{R}^d))^{\otimes n}$
$\mathcal{A}$	$:= (1-B_L(r))\frac{d^2}{dr^2} + \left((d-1)\frac{1-B_N(r)}{r} - cr\right)\frac{d}{dr}$
a.s.	almost sure(ly)
$A_1,\ldots,A_n$	Borel subsets of $\mathrm{Diff}(\mathbb{R}^d)$
$A_1,\ldots,A_n$	Borel subsets of $\mathbb{R}^d$

Index of Notation

A	linear mapping from $\mathbb{R}^d$ to $\mathbb{R}^d$
A, A_n	constant
A_t	process of locally bounded variation
a_i	numbers smaller than 1
$\text{Ap}(E, \tilde{E})$	aperture between E and $\tilde{E}$
$\hat{a}$	real number $\geq \hat{b}$
$b(x,s) = b^i(x,s)_{1 \leq i \leq n}$	drift part of local characteristic
$(B^{m,\delta}, B^{m',\delta'})$	function class see page 11
$(B_b^{m,\delta}, B_b^{m',\delta'})$	function class see page 11
$(B_{ub}^{m,\delta}, B_{ub}^{m',\delta'})$	function class see page 11
$B^{m,\delta}$	function class see page 11
$B_b^{m',\delta'}$	function class see page 11
$B_{ub}^{m,\delta}$	function class see page 11
$\mathcal{B}(\cdot)$	Borel σ-field
$b = b(x)$	isotropic covariance tensor
B_L	longitundinal correlation function
B_N	normal correlation function
$\bar{b}$	diffusion matrix of (x_t, y_t)
$\mathcal{B}_\Omega$	σ-algebra in Ω
$B_1, \ldots, B_m$	partition of Ω
$\mathcal{B}(\cdot)$	subset of $\mathbb{R}^2$ - limit shape
B, B_n	constant
b_i	numbers smaller than 1
$\hat{b}$	real number $\geq \hat{a}$
$C^m(\mathbb{D} : \mathbb{R}^n)$	m times continuously differentiable functions from $\mathbb{D}$ to $\mathbb{R}^m$
$C_b^m(\mathbb{D} : \mathbb{R}^n)$	$:= \{f \in C^m(\mathbb{D} : \mathbb{R}^n) : \|f\|_m < \infty\}$
$C^{m,\delta}(\mathbb{D} : \mathbb{R}^n)$	$:= \{f \in C^m(\mathbb{D} : \mathbb{R}^n) : D_x^\alpha f \text{ is } \delta - \text{Hölder continuous}\}$
$C^{m,\delta}$	function class - see page 8
$C_b^{m,\delta}$	function class - see page 8
$C_{ub}^{m,\delta}$	function class - see page 8

Index of Notation

C^m	function class - see page 8
C_b^m	function class - see page 8
C_{ub}^m	function class - see page 8
$\tilde{C}^m(\mathbb{D}:\mathbb{R}^n)$	C^m functions from $\mathbb{D} \times \mathbb{D}$ to $\mathbb{R}^m$
$\tilde{C}_b^m$	$:= \{g \colon \mathbb{D} \times \mathbb{D} \to \mathbb{R}^n : \|g\|_m^\sim < \infty\}$
$\tilde{C}_b^{m,\delta}$	$:= \{g \colon \mathbb{D} \times \mathbb{D} \to \mathbb{R}^n : \|g\|_{m+\delta}^\sim < \infty\}$
$\tilde{C}^{m,\delta}$	function class - see page 10
$\tilde{C}_b^{m,\delta}$	function class - see page 10
$\tilde{C}_{ub}^{m,\delta}$	function class - see page 10
$\tilde{C}^m$	function class - see page 10
$\tilde{C}_b^m$	function class - see page 10
$\tilde{C}_{ub}^m$	function class - see page 10
$C^{m,\delta}$-local martingale	see Theorem 1.1.1
$C^{m,\delta}$-semimartingale	see Theorem 1.1.1
$C^{m,\delta}$-valued Brownian motion	see page 11
C	generic notation for a constant
C_b^2	cont. funct. with bounded derivatives. up to order 2
C_0^∞	smooth functions with compact support
$c_1, c_2, \ldots$	generic constants - new numbering per chapter
$C_1, C_2, \ldots$	generic constants - new numbering per chapter
$C(\mathbb{R}^d \times \{1,\ldots,d\} : \mathbb{R})$	$:= \{f \colon \mathbb{R}^d \times \{1,\ldots,d\} \to \mathbb{R} : f \text{ is continuous}\}$
c	positive constant - drift of an IOUF
cf.	confer, see also
$C(d)$	d-depending constant
$\mathcal{C}_R$	$:= \{\gamma : \mathrm{diam}(\gamma) \geq 1, \gamma \subset K_{2R}(0)\}$
$\mathcal{C}_R^*$	set of all large curves γ with $\gamma \cap \partial K_R(0) \neq \emptyset$
C_q	constant in the Burkholder-Davies-Gundy inequality
$\bar{c}, \bar{c}_1, \bar{c}_2, \ldots$	constants
$\bar{C}_1, \bar{C}_2, \ldots$	constants
$\tilde{c}$	$:= 2\beta_L + 2d^6 \max_{i,l,k,n} \sup_{z \in \mathbb{R}^d} \partial_k \partial_n b^{i,l}(z)$

Index of Notation

$\hat{C}$	real number ≥ 1
d	natural number, dimension of the state space
$\mathbb{D}$	domain in $\mathbb{R}^d$
D_x^α	spatial differential operator coming from the multi-index α
$d(\phi,\psi)$	$:= \rho(\phi,\psi) + \rho(\phi^{-1},\psi^{-1})$
$d_k(\phi,\psi)$	$:= \sum_{|\alpha|\leq k}\rho(D^\alpha\phi, D^\alpha\psi) + \sum_{|\alpha|\leq k}\rho(D^\alpha\phi^{-1}, D^\alpha\psi^{-1})$
DS	dynamical system
$d_i(\omega)$	multiplicity of ith Lyapunov exponent
d_t	$:= \text{diam}(\gamma_t)$
d_ν	distance along $\nu_{(\omega,z),m}^\lambda$ for $m \in \mathbb{N}$
e	Euler's number
$\mathbb{E}[\cdot]$	expectation operator
E	Borel subset of $\mathbb{R}^{nd}$
$E_1(\omega),\ldots,E_{p(\omega)}(\omega)$	splitting in the MET for linear RDS
$E_i(\omega,x))$	Oseledec splitting in the MET
E_n	$n \times n$ unity matrix
e.g.	exempli gratia, for example
a.e.	almost every
$K_r(0)$	closed r-ball centered at $x \in \mathbb{R}^d$
$E_{k,l}$	maximal $\frac{1}{k}$-separated set in $K_l(0)$
$E_0(\omega,z)$	$:= \oplus_{\lambda_i<\lambda}E_i(\omega,z)$
$E_n(\omega,z)$	$:= D_z\varphi_n(\cdot,\omega)E_0(\omega,z)$
$E, \tilde{E}$	vector space
$\mathcal{F}$	generic notation for a σ-field
$\bar{\mathcal{F}}$	$\mathbb{P}$-completion of $\mathcal{F}$
f	generic notation for an $\mathbb{R}^n$-valued function
$F = F(t,x,\omega)$	generic notation for a semimartingale field
$\hat{F} = \hat{F}(t,x,\omega)$	backward version of $F(t,x,\omega)$
$\{f_t : t \geq 0\}$	predictable process with values in $\mathbb{D}$
f_t^n	approximating sequence to get the stochastic integral

Index of Notation

$\{\mathcal{F}_s^t : s, t \in \mathbb{R}, s \leq t\}$	generic notation for a two-parameter filtration
$\mathcal{F}_{-\infty}^t$	$:= \bigvee_{s \leq t} \mathcal{F}_s^t$
$\hat{\mathcal{F}}_s^\infty$	$:= \bigvee_{t \geq s} \mathcal{F}_s^t$
$\tilde{\mathcal{F}}$	σ-field
$\{\tilde{\mathcal{F}}_s^t : s, t \in \mathbb{R}, s \leq t\}$	two-parameter filtration
f	endomorphism of a probability space
$F_1(t), F_2(t)$	events in Ω
F_n	$:= \{\exists i \in [\lfloor\sqrt{n}\rfloor, \infty] \cap \mathbb{N} : d_i < 1\}$
F	event in Ω
$F : \Omega \times \mathbb{R}^d \to C^{2,1}(\mathbb{R}^d, \mathbb{R}^d):$	$(\omega, z) \mapsto F_{(\omega,z)} = \varphi_1(z + \cdot, \omega) - \varphi_1(z, \omega)$
$F_{(\omega,z)}^n$	$:= F_{\tau^{n-1}(\omega,z)} \circ \ldots \circ F_{\tau(\omega,z)} \circ F_{(\omega,z)}$
	$= \varphi_n(z + \cdot, \omega) - \varphi_n(z, \omega)$ for $n \in \mathbb{N}$
g	generic notation for a $\mathbb{R}^n$-valued function
$G(t, x, y, \omega)$	function : $[0, T] \times \mathbb{D} \times \mathbb{D} \times \Omega \to \mathbb{R}^n$
G	group of homeomorphisms of $\mathbb{R}^d$
G^k	group of C^k-diffeomorphisms of $\mathbb{R}^d$
$\mathcal{G}$	sub-σ-algebra of $\mathcal{F}$
$\tilde{\mathcal{G}}$	sub-σ-algebra of $\mathcal{F}$
G', G''	functions in the chasing balls lemma
$\bar{g}(R, \tilde{R}, t)$	$:= \sup_{\gamma \in \mathcal{C}_{\tilde{R}}} \left\{ \mathbb{E}\left[\tau^R(\gamma, tv)\right] \right\}$
h	generic notation for a $\mathbb{R}^n$-valued function
$h(x, i)$	element of the RKHS
$H_\mathbb{P}(\xi \mid \mathcal{G})$	conditional entropy of ξ given $\mathcal{G}$
$h_\mathbb{P}^\mathcal{G}(\xi, f)$	entropy of f with respect to ξ given $\mathcal{G}$
$h_\mathbb{P}^\mathcal{G}(f)$	entropy of f given $\mathcal{G}$
$h_\mathbb{P}(f)$	entropy of f
$h_\mu(\phi, \xi)$	$:= \lim_{n \to \infty} \frac{1}{n} \int H_\mu \left(\bigvee_{i=0}^{n-1} \phi_{0,i}(\cdot, \omega)^{-1} \xi \right) d\mathbb{P}(\omega)$
$h_\mu(\phi)$	metric entropy of ϕ
h	L^2-control function
$\tilde{g}$	diffusion term in SDE

$\mathcal{H}$	RKHS of Φ
$h_{(\omega,z)}$	function $K_{\alpha(\omega,z)}(0) \subset \oplus_{\lambda_i<\lambda} E_i(\omega,z) \to (\oplus_{\lambda_i<\lambda} E_i(\omega,z))^\perp$
$H_0(\omega,z)$	$:= E_0(\omega,z)^\perp$
$H_n(\omega,z)$	$:= D_z\varphi_n(\cdot,\omega)H_0(\omega,z)$
H	Hilbert space
i	natural number
iff	if and only if
IBF	isotropic Brownian flow
IOUF	isotropic Ornstein-Uhlenbeck flow
$\mathcal{I}$	isomorphism
i.i.d	independend, identically distributed
i.e.	this is
$\mathcal{I}$	symmetric ideal in $V^\otimes$
j	natural number
$\mathbb{K}$	compact subset of $\mathbb{R}^n$
k	natural number
K	generic notation for a constant
K	expansion speed for the derivatives
$k_1, k_2, \ldots$	generic notation for constants - new numbering per chapter
$K_1, K_2, \ldots$	generic notation for constants - new numbering per chapter
$k(x,i,y,j)$	kernel of the covariance operator of $\mathcal{N}$
$K_\epsilon(X)$	$:= \{y \in \mathbb{R}^d : \text{dist}(y,X) \leq \epsilon\}$ closed ϵ-neighbourhood of X
$K_r(x)$	$:= \{y \in \mathbb{R}^d : \text{dist}(y,x) \leq r\}$ closed r-ball centered at x
l	fuction from $\mathbb{N}$ to $\mathbb{N}$
k	real constant
l	natural number
$\mathcal{L}$	law of a random variable
$L^{(n)}$	generator of $(\Phi_t(x^{(1)}), \ldots, \Phi_t(x^{(n)}))$
$\bar{L}$	generator of ρ_t^{xy}
$\mathcal{L}$	$:= -c\sum_{i=1}^d x_i \frac{\partial}{\partial x_i} + \frac{1}{2}\Delta$.

INDEX OF NOTATION 201

$\mathcal{L}_d$	$:= -c\sum_{i=1}^d x_i \frac{\partial}{\partial x_i} + \sum_{i,j=1}^d (\delta_{ij} - b_{ij}(x))\frac{\partial^2}{\partial x_i \partial x_j}$
$L_k(n, \omega, i)$	$:= \sup_{z \in K_{\frac{1}{k}}(x_i)} \|D_z \phi_n\|$
$l(n, \omega, x)$	measurable function $l : \Gamma \times \mathbb{N} \to (0, \infty)$
l'	real number
m	natural number
$M = M(t, x, \omega)$	generic notation for the local martingale part of $F(t, x, \omega)$
MDS	metric dynamical system
MET	multiplicative ergodic theorem
M	manifold
m	$:= \mathcal{L}(\phi_{0,1}(\cdot))$ law of unit step discretization of an IOUF
M_t	$:= 2\int_0^t \sum_{i,j,k} \partial_k M^i (ds, x_s) \frac{\partial_j \phi_s^k(x) \partial_j \phi_s^i(x)}{\|D\phi_s(x)\|^2}$
$\tilde{M}_t$	special continuous martingale
$\check{M}_t$	$= \check{M}_t^{(ij)} := \int_0^t X_s^{(ij)q-1} dVII_t - \int_0^t X_s^{(ij)q-1} dVIII_t$
$\mathbb{N}$	the set of natural numbers starting with 1
n	natural number
$N = N(t, x, \omega)$	martingale field
$\tilde{\mathcal{N}}$	Gaussian measure
$\mathcal{N}(\cdot, \cdot)$	normal distribution
$\mathcal{N}$	Gaussian measure
N_δ	natural number depending on δ
O	orthogonal matrix
$O(d)$	group of orthogonal $d \times d$ matrices
O^*	adjoint of O
$\mathbb{P}$	probability measure
$P_{s,t}^{(n)}(\cdot, \cdot)$	Markov semigroup
$p(\omega)$	number of Lyapunov exponents
$\tilde{\mathbb{P}}$	probability measure
P_t	Markov semigroup
$p, p_1, \ldots, p_3$	reals in $(0, 1)$, strictly positive probabilities
P	point in $\mathbb{R}^d$ far away from the origin

$P_{\beta,x,j}(s)$	finite sum of products of drivatives of $\phi_s(x)$
q	real number ≥ 1
$Q = (q_1, q_2)$	point in $K_R(P) \subset \mathbb{R}^2$
$\tilde{Q}$	point in U_Q
$\mathbb{R}$	the real numbers
r	real number
RDS	random dynamical system
$r^{(\epsilon)}$	positive constant
$\bar{r}$	positive constant
RKHS	reproducing kernel Hilbert space
$\mathbb{R}^{2d}_\times$	$:= \mathbb{R}^{2d} \setminus \{z \in \mathbb{R}^{2d} : z_i = z_{d+i} \forall i = 1, \ldots, d\}$
$r_i, i = 1, 2$	random distances in the ball chasing lemma
$R, \tilde{R}$	positive real numbers, radii of certain balls in $\mathbb{R}^d$
r_t	$:= \mathrm{dist}(\gamma_t, P)$
s	real number - time
SDE	stochastic differential equation
$S_n^l(\omega, z)$	$:= T_n^l(\omega, z)|_{E_n(\omega, z)}$
t	real number - time
T	real number - time horizon
$T(t)$	linear bundle RDS coming from the differential as RDS
$T_x M$	tangent space at x to M
TM	tangent bundle of M
T	endomorphism of V
$T^{\wedge k}$	kth exteriour power of T
$T^\wedge : V^\wedge \to V^\wedge, T^\wedge$	$:= \mathrm{id}_\mathbb{R} \oplus T \oplus T^{\wedge 2} \ldots T^{\wedge n}$
$(t_k)_{k \in \mathbb{N}}$	sequence of random times
$T(\gamma, \epsilon) > 0$	random time
t_u^n	$:= \frac{n\epsilon}{2} \left(\sup_{z \in U_Q^{n,\epsilon}} (\|V_1(z)\| \vee \|V_2(z)\|) \right)^{-1}$
t_j	$:= \inf \{t \in \mathbb{R} : t \geq t_{j-1} + 1 + T : \gamma_t \cap K_R(P) \neq \emptyset, \mathrm{diam}(\gamma_t) \geq 1\}$
$\tilde{t}_m^{(n)}(\delta)$	positive constant

Index of Notation

$T_{(\omega,z)}$	$:= D_z\varphi_1(\cdot,\omega)$
$T^n_{(\omega,z)}$	$:= T_{\tau^{n-1}(\omega,z)} \circ \ldots \circ T_{\tau(\omega,z)} \circ T_{(\omega,z)} = D_z\varphi_n(\cdot,\omega)$ for $n \in \mathbb{N}$
$T^l_n(\omega,z)$	$:= D_{\varphi_n(z,\omega)}\varphi_{n+l}(\cdot,\theta^n\omega)$
$(T_i(x) : x \in X, i \in \mathbb{N})$	be a family of bounded linear operators $T_i(x)\colon H \to H$
u	real number - time
$U = U(x)$	isotropic Brownian field
U_Q	open superset of $\{Q\}$
$U_Q^{n,\epsilon}$	$:= \,]q_1 - n\epsilon, q_1 + n\epsilon[\,\times\,]q_2 - n\epsilon, q_2 + n\epsilon[$
$\tilde{U}_Q^n$	$:= \,]q_1 - \frac{n\epsilon}{2}, q_1 + \frac{n\epsilon}{2}[\,\times\,]q_2 - \frac{n\epsilon}{2}, q_2 + \frac{n\epsilon}{2}[$
$\hat{U}_Q$	$:= Z^{-1}\,(\,]-7,7[^2)$
$U^l_n(\omega,z)$	$:= T^l_n(\omega,z)|_{H_n(\omega,z)}$
$U^l_n(\omega,z)$	$:= T^l_n(\omega,z)|_{H_n(\omega,z)}$
$V = V(t,x,\omega)$	generic notation for the bounded variation part of $F(t,x,\omega)$
v	vector in $\mathbb{R}^d$
$V_i(x,\omega)$	Oseledec splitting for IBFs
$V_i(x)$	orthonormal Hilbert base for the RKHS
V^U	isotropic Brownian field with potential U
V	finite-dimensional vector space
$V^\otimes$	tensor algebra over V
$V^\wedge$	outer algebra over V
$V^{\wedge k}$	$:= V^{\otimes k} \cap V^\wedge$
$V_1^i(.)$	$:= \int b^{ij}(.-y)d\delta_Q \otimes \delta_1(y,j) = b^{i,1}(.-Q) : i = 1,2$
$V_2^i(.)$	$:= \int b^{ij}(.-y)d\delta_Q \otimes \delta_2(y,j) = b^{i,2}(.-Q) : i = 1,2$
V	$\mathcal{H}$-simple control
w_1	$:= V_1(Q) = b^{\cdot 1}(0) = \begin{pmatrix} 1 \\ 0 \end{pmatrix}$
w_2	$:= V_2(Q) = b^{\cdot 2}(0) = \begin{pmatrix} 0 \\ 1 \end{pmatrix}$
$(W_t : t \geq 0)$	generic notation for a Brownian motion
$(W_t^i : t \geq 0), i = 1,2$	Brownian motions

$(W_t^* : t \geq 0)$	running maximum for Brownian motion
$\mathcal{W}_t(\gamma)$	$:= \bigcup_{0 \leq s \leq t} \gamma_s$
$\mathcal{W}_t^R(\gamma)$	$:= \left\{ x \in \mathbb{R}^d : \text{dist}\,(x, \mathcal{W}_t(\gamma)) \leq R \right\}$
w.r.t.	with respect to, with reference to
w.l.o.g.	without loss of generality
w.p.p.	with positive probability
x	point in $\mathbb{R}^d$
$x^{(1)}, x^{(2)}, \ldots$	points in $\mathbb{R}^d$
$(X_t : t \geq 0)$	stochastic process - solution to an SDE
$(X_t)_{t=1,2,\ldots}$	discrete time Markov chain on $\mathbb{R}^d$
$\mathbb{X}$	compact metric space
$\hat{x}(x)$	$:= g^{-1} \circ f(x)$
$x^{(1)}$	point in $\mathbb{R}^d$
$(X_t : t > 0)$	family of integrable random variables
$X_t^{(ij)}$	$:= \frac{\partial_j \phi_t^i(x)}{\|D\phi_t(x)\|} - \frac{\partial_j \phi_t^i(y)}{\|D\phi_t(y)\|}$
$X_t^{(ij)q}$	$:= \left(X_t^{(ij)}\right)^q$
(X, d)	metric space
y	point in $\mathbb{R}^d$
$(Y, \mathcal{Y})$	measurable space
$y^{(1)}$	point in $\mathbb{R}^d$
$\tilde{y}$	point in $\mathbb{R}^2$
Y_t	$:= X_t - \mathbb{E}\left[X_t\right]$
z	point in $\mathbb{R}^d$
z	real number
z_1, z_2	components of $z \in \mathbb{R}^2$
$z_t(h)$	solution of control problem driven by h
$Z := (Z_1, Z_2) : \tilde{U}_Q^{102} \to\,]-51, 51[$	coordinate system
$\alpha = (\alpha_1, \ldots, \alpha_n)$	multi-index
$\alpha = \alpha(\omega, z)$	real function on $\text{Diff}(\mathbb{R}^d)^{\mathbb{Z}} \times \mathbb{R}^d$
$\alpha = \alpha(\omega)$	random variable

Index of Notation

β_L	$:= -B_L''(0)$
β_N	$:= -B_N''(0)$
β	real number > 1
$\beta = \beta(\omega, z)$	real function on $\mathrm{Diff}(\mathbb{R}^d)^{\mathbb{Z}} \times \mathbb{R}^d$
$\beta > 0$	Hölder exponent
β	multi-index
γ	original set for the limit shape theorem
γ_t	$:= \Phi_t(\gamma)$
$\check{\gamma}, \hat{\gamma}, \hat{\hat{\gamma}}, \bar{\gamma}, \gamma^{(t)}$	large subsets of $\mathbb{R}^d$
Γ	boundary of the unit ball in $\mathbb{R}^2$
$\gamma^{(i)}$	random large subsets at τ_i^R
$\gamma = \gamma(\omega, z), \gamma' = \gamma'(\omega, z)$	real functions on $\mathrm{Diff}(\mathbb{R}^d)^{\mathbb{Z}} \times \mathbb{R}^d$
Γ	set of full measure
γ	angle between two vector spaces
$\Gamma_\epsilon^{l'}$	$:= \{(\omega, z) \in \Gamma \colon l(\omega, z, 0) \leq l'\}$
γ	multi-index
δ	positive real number - sometimes Hölder exponent
Δ	a.s. event in the MET for smooth cocycles
Δ	Laplacian
δ_{ij}	Kronecker's delta
$\delta_{(x,i)}$	evaluation functional
$\delta_1, \delta_2, \ldots$	sumable real sequence
$\delta_1(A), \ldots, \delta_d(A)$	singular values of A
Δ	upper entropy dimension of Ξ
$\Delta_{\hat{a},\hat{b},\hat{C}}$	set in X
δ	multi-index
ϵ	positive real number
$\epsilon_1, \epsilon_2, \ldots$	sequence of positive reals
$\tilde{\epsilon}$	positive real number
η	countable partition of Ω

η	element of $E_n(\omega, z)^\perp$
θ_t	shift-mapping in a DS, MDS or RDS
θ	positive constant in $(0, 1)$
ι	function on $[0, \infty)$ given by $\iota : x \mapsto x \log x$
$(\kappa_m^{(i)})$	$m \in \mathbb{N} : i = 1, 2, \ldots$ tail constants
κ_q	q-depending constant
λ, λ'	constant
λ_i	Lyapunov exponents of an IOUF
$\Lambda, \Lambda_0, \Lambda_1, \ldots$	constants
$\tilde{\lambda}$	largest Lyapunov exponent smaller than λ.
μ_i	Lyapunov exponents of an IBF
μ	invariant measure for a RDS
$\nu_{(\omega,z)}^\lambda$	$:= \{x \in K_{\alpha(\omega,z)}(0) : \forall n \geq 0 : \|\varphi_n(z+x, \omega) - \varphi_n(z, \omega)\| \leq \beta(\omega, z)e^{\lambda n}\}$
$\nu_{(\omega,z),n}^\lambda$	$:= \{x \in K_{\alpha_n(\omega,z)}(0) :$ $\forall m \geq 0 : \|\varphi_m(\varphi_n(z,\omega)+x, \theta^n\omega) - \varphi_m(\varphi_n(z,\omega), \theta^n\omega)\| \leq \beta_n(\omega, z)e^{\lambda m}\}$
d_ν	distance along $\nu_{(\omega,z),m}^\lambda$ for $m \in \mathbb{N}$
$\nu_{(\omega,z),n}^{\lambda,g}$	$:= \{x \in \mathbb{R}^d : \lim_{m \to \infty} \frac{1}{m} \log \|\varphi_m(\varphi_n(z,\omega)+x, \theta^n\omega) - \varphi_m(\varphi_n(z,\omega), \theta^n\omega)\| \leq \lambda\}$
ξ	countable partition of Ω
$(\xi_i)_{i \in \mathbb{N}}$	sequence of measurable partitions of Ω
ξ_x	element of the partition ξ that contains x
Ξ	$:= f^{-1} \circ \Psi \circ f$
$\xi_j, j \in \mathbb{N}$	discrete time supermartingale
$\xi_j^{(C_{10}, \delta)}$	special choice for ξ_j
Ξ	bounded subset of $\mathbb{R}^d$
ξ	generic element of $K_{\alpha_n(\omega,z)}(0) \cap E_n(\omega, z)$
ξ_n	sequence in $E_0((\omega, x)_n)$
π	projection
π	$\pi = 3.141\ldots$
$\rho(\phi, \psi)$	$:= \sum_{N=1}^\infty \frac{\sup_{|x| \leq N} |\phi(x) - \psi(x)|}{2^N(1+\sup_{|x| \leq N} |\phi(x) - \psi(x)|)}$
ρ_t^{xy}	$:= \|x_t - y_t\|$

$\rho(f,g)$	$:= \sum_{n=0}^{\infty} \frac{1}{2^n} \frac{\rho_n(f,g)}{1+\rho_n(f,g)}$
$\rho_n(f,g)$	$:= \max_{i=1,\ldots,d} \sup_{|x|\leq n} |f(x,i) - g(x,i)|$
ρ	metric on $\mathbb{X}$
ρ	positive constant in the chasing ball lemma
ρ	$:= \mathbb{P} \otimes \mu$
$\bar{\sigma}$	positive constant
$\sigma, \sigma_1, \sigma_2, \ldots$	constants
τ_t	skew product shift of RDS
$\tau^R(\gamma, P)$	$:= \inf\{t > 0 : \text{dist}(\gamma_t, P) \leq R, d_t \geq 1\}$
$\tau_i^R, i = 1, 2, \ldots$	sequence of stopping times
$\tilde{\tau}$	$= \tilde{\tau}^R(\gamma, P) := \tilde{\tau}(P) := \inf_{t>0} \{K_R(P) \subset \bigcup_{0\leq s\leq t} \gamma_s\}$ sweeping time
$\tilde{\tau}_Q$	$:= \inf_{t>0} \{U_Q \subset \bigcup_{0\leq s\leq t} \gamma_s\}$
τ	stopping time
$\phi_{s,t}(x,\omega)$	stochastic flow
$\Phi_{s,t}(x,\omega)$	IBF
$\phi_{s,t}(x,\omega)$	IOUF
$\phi_{s,t}(x,\omega)$	IBF or IOUF
ϕ	random function on $\mathbb{X}$
$\tilde{\Phi}_{s,t}(x,\omega)$	independent copy of Φ
φ	random dynamical system
χ_i	skeleton in $\mathbb{X}$
Ψ	global measurable trivialization
$\psi = \psi_t(x)$	solution to a control problem
Ψ	unit ball version of IOUF
$\psi_t^{(V)}(x)$	solution to the control problem driven by V
$\Psi(.,.)$	a continuous mapping from $[0,T] \times \mathbb{R}^2$ to $\mathbb{R}^2$
$\tilde{\psi}(s,u)$	$:= \psi_s^{(V)}(\tilde{\gamma}(u))$
$\tilde{\Psi}(s,u)$	$:= \Psi(s, \tilde{\gamma}(u))$
$\bar{\psi}(s,u)$	$:= \psi_s^{(V)}(\bar{\gamma}(u))$
$\bar{\Psi}(s,u)$	$:= \Psi(s, \bar{\gamma}(u))$

$\psi_t(x)$	$:= \log \lVert D\phi_t(x) \rVert$
Ω	base set of the generic probability space
ω	element of Ω
$\tilde{\Omega}$	sure subset of Ω
$\tilde{\Omega}$	base set of a probability space
$\tilde{\omega}$	element of $\tilde{\Omega}$
$\Omega_{k,l}$	subset of Ω

Index of Notation

Die VDM Verlagsservicegesellschaft sucht für wissenschaftliche Verlage abgeschlossene und herausragende

Dissertationen, Habilitationen, Diplomarbeiten, Master Theses, Magisterarbeiten usw.

für die kostenlose Publikation als Fachbuch.

Sie verfügen über eine Arbeit, die hohen inhaltlichen und formalen Ansprüchen genügt, und haben Interesse an einer honorarvergüteten Publikation?

Dann senden Sie bitte erste Informationen über sich und Ihre Arbeit per Email an *info@vdm-vsg.de*.

Sie erhalten kurzfristig unser Feedback!

VDM Verlagsservicegesellschaft mbH
Dudweiler Landstr. 99   Telefon +49 681 3720 174
D - 66123 Saarbrücken   Fax   +49 681 3720 1749
www.vdm-vsg.de

Die VDM Verlagsservicegesellschaft mbH vertritt

Printed by Books on Demand GmbH, Norderstedt / Germany